THE BATS OF BRITISH COLUMBIA

BATS

of British Columbia

David W. Nagorsen
Vertebrate Zoology Unit
Royal British Columbia Museum

R. Mark Brigham
Department of Biology
University of Regina

Illustrations by Michael Hames

Volume 1
The Mammals of British Columbia

UBC PRESS / VANCOUVER

ISBN 0-7748-0482-3
ISSN 1188-5114 (RBCM Handbook series)

Canadian Cataloguing in Publication Data

Nagorsen, David W.
  The Bats of British Columbia

  (Royal British Columbia Museum handbook,
ISSN 1188-5114)
  Co-published by the Royal British Columbia
  Museum.
  Includes bibliographical references and index.
  ISBN 0-7748-0482-3

  1. Bats—British Columbia.   I. Brigham,
Robert Mark, 1960–   II. Royal British Columbia
Museum.   III. Title.   IV. Series.
QL737.C5N33 1993   599.4'09711   C93-091797-9

The colour map, "Biogeoclimatic Zones of British Columbia", was
produced and supplied by the Forest Science Research Branch, Ministry of
Forests.

Cover photographs by Merlin D. Tuttle, Bat Conservation International
Printed in Canada on acid-free paper

Production of this book was funded, in part, by the Ministry of Environment,
Lands and Parks and the Ministry of Forests, through the Corporate Resources
Inventory Initiative.

UBC Press
University of British Columbia
6344 Memorial Road
Vancouver, BC V6T 1Z2
(604) 822-3259
Fax: (604) 822-6083

CONTENTS

PREFACE

For nearly four decades the Royal British Columbia Museum's Handbook 11, *The Mammals of British Columbia* by Ian McTaggart Cowan and Charles J. Guiguet, has been the standard reference on the province's mammals. Last revised in 1965, this handbook is now out of print. Changes in mammalian taxonomy and nomenclature, new additions to the province's mammalian fauna, range extensions, and an explosion in ecological research during the past three decades have made *The Mammals of British Columbia* out of date. With the growing demand for information on small mammals from government wildlife and forestry agencies, environmental groups and the general public, the time is appropriate for a revised handbook on the province's mammals.

Rather than summarizing the province's 143 native and introduced species in a large single volume, the museum will produce a series of six handbooks. This volume on bats is the first in the new series. The reader will note several departures in style and format from the original mammal handbook. Although the new handbooks will certainly be useful to mammalogists and other biologists they are directed especially towards a more general audience. Emphasis is on identification, natural history, distribution and conservation; taxonomy is limited to a listing of subspecies.

The choice of bats for the first handbook may appear odd to many readers. It reflects the authors' long fascination with this group and the high interest in bats demonstrated by the general public. Except for marine mammals, the Royal British Columbia Museum probably receives more general inquiries about bats than any other group of mammals. With the growing interest in preserving biodiversity in the province, more attention is being directed by government agencies towards the various species of small mammals. Nowhere is this more evident than the bats. Eight species now appear on the Red and Blue lists of potentially endangered and threatened species prepared by the provincial Ministry of Environment.

British Columbia has the greatest diversity of bats of all Canadian provinces. Eight of the species found here occur nowhere else in Canada. Clearly the province has an obligation to ensure the survival of these mammals. Unfortunately the biology of many bats in British Columbia is poorly known. Comprehensive field studies applying modern research techniques have been carried out only in the Okanagan Valley. We hope that this book, besides summarizing existing information, also identifies our lack of knowledge and stimulates more research on this fascinating group of mammals.

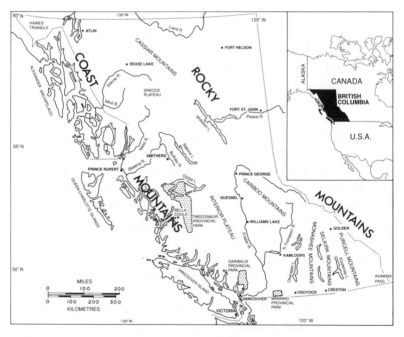

Figure 1. General geographic features of British Columbia.

GENERAL BIOLOGY

No other group of mammals is as shrouded in fear and misinformation as bats. Most people have never seen a bat up close and their perception of bats is based largely on myth or superstition rather than personal knowledge or scientific fact. In reality bats are fascinating mammals that display some remarkable adaptations for their unique lifestyle

Bats are the only mammals capable of sustained, flapping flight, a characteristic that they share with birds and the extinct pterosaurs. The success of bats in terms of great diversity of species, nearly worldwide distribution, huge populations of some species and diverse methods of capturing prey are all linked to their ability to fly. Two other important adaptations of bats are a sonar or echolocation system that enables most species to navigate and find food in total darkness, and an internal thermostat that enables many bats to lower their body temperature and become inactive or torpid at cool temperatures, thus avoiding the energy demands of maintaining a high body temperature.

The origin of bats remains obscure and controversial. The oldest fossil of a bat was found in North America and is thought to be 50 to 60 million years old. Fossil bats differ little from modern-day insectivorous bats in their bone structure; unfortunately, scientists have not yet found a fossil demonstrating the transition from a land-dwelling mammal to a modern bat. Nevertheless, most biologists believe that bats evolved from an ancestral relative of the insectivores (Insectivora), a primitive group of mammals that includes the shrews and moles. The shrew-like ancestor of bats was probably a tree-dwelling glider and had a fexible membrane of skin that extended along the sides of the body and between the fingers of the forelimb not unlike the gliding membrane of the colugo or flying lemur (Dermoptera). Although the evidence is scanty, some scientists take this scenario a step further and speculate that ancestral bats were capable of detecting insects using echolocation. If this were true, primitive bats could have successfully operated at night thereby avoiding competititon with well-established, diurnal, insectivorous birds.

Whatever the evolutionary scenario, bats are a successful group. They are sufficiently distinct from other mammals that taxonomists classify them in their own mammalian order, the Chiroptera, which means "hand-wing". The living species of Chiroptera are subdivided into two major groups: the Microchiroptera comprising about 900 species that typically use echolocation and the Megachiroptera, some of which are referred to as flying foxes, a group that contains about 165 species. The Megachiroptera are restricted to the Old World tropics whereas the Microchiroptera are found worldwide except in polar regions.

Recently there has been a vigorous debate in the scientific community about the ancestors of the Megachiroptera. Traditionally both groups were thought to have evolved from the same ancestral insectivore. However, an Australian biologist recently discovered that flying foxes, primates (including humans) and the colugos or flying lemurs share a unique attribute in the nerve structure of their brains. This raises some provocative questions: do the Microchiroptera and Megachiroptera stem from a common ancestor or did the Megachiroptera evolve from a primate ancestor? If the latter is true, then the flying foxes might be more appropriately described as flying monkeys! More research is needed before these interesting problems can be resolved. However, it is clear that despite the popular perception of bats as "flying mice" they show no close relationship to rodents.

Bats inhabit every continent except Antarctica and can be found in virtually every type of habitat from desert to forest. Most species live in the tropics and it is in that region where bats show the greatest variety of forms and life-styles. Although the majority are small, weighing less than fifty grams, bats come in many sizes. The largest, the flying foxes of Asia, can attain a body mass of more than one kilogram and a wingspan of two metres. Despite their formidable size, the flying foxes are harmless fruit-eaters. The smallest bat in the world is the rare Bumblebee Bat (*Craseonycteris thonglongyai*) of Thailand, which weighs about two grams and has a wingspan of fifteen centimetres. About three-fourths of the world's bats consume insects and other invertebrates, but there are species in the tropics that prey on vertebrates such as frogs, reptiles, fish and small mammals including other bats. The most specialized of all the animal eaters are the three species of New World vampire bats that feed exclusively on the blood of birds and mammals. Other tropical bats are strictly vegetarian, feeding on the fruit, pollen and nectar of plants.

We have none of these exotic tropical forms in British Columbia. All of our 16 native species belong to the family Vespertilionidae, a large family that ranges farther into temperate regions than any other bat family. The vespertilionid bats are insect eaters; nevertheless, they are a fascinating group because many have evolved special adaptations to cope with the severe conditions of temperate winters. The general biology of vespertilionids is discussed in the following sections. This is a brief overview. If the reader is interested in obtaining more detailed information on bats, consult the general books or selected references on bat biology listed in the References section.

Form and Structure

All bats have forelimbs modified for flight. The bat wing consists of a thin, double-layered membrane of skin stretched over the arms, hands and fingers (Figure 2). Another flight membrane, the tail membrane, extends between the hind legs and tail. These flight membranes are elastic and flexible, yet relatively tough and resistant to tearing, somewhat like a thin sheet of plastic. The wings contain muscles and blood vessels that are important for thermoregulation. Most of the surface of the wings and tail membranes is devoid of hair to reduce drag when the bat is in flight, although the regions around the upper arm and base of the tail are often furred. The bones of the arm, hand and fingers provide the internal support for the wing. They conform to the usual mammalian design of an upper arm, forearm, wrist and hand with a thumb and four fingers, but the bones of the hand and fingers are greatly elongated. Only the thumb has a claw. When a bat is resting, it draws its finger bones together and holds its arms against its body, folding the wing membrane to prevent injury. In flight, a bat stretches out its arms and fingers with the wing membranes taut over the bones.

The inside portion of the wing, between the body and the fifth

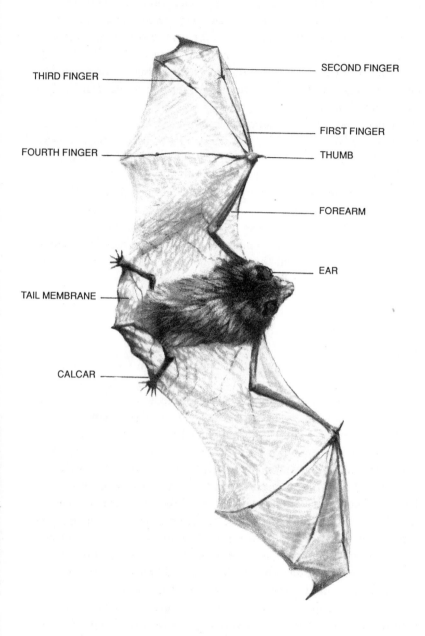

THIRD FINGER

SECOND FINGER

FIRST FINGER

FOURTH FINGER

THUMB

FOREARM

EAR

TAIL MEMBRANE

CALCAR

Figure 2. General external features of a vespertilionid bat.

finger, produces most of the lift; movement of the wing tip, from the fifth finger to the edge, generates most of the thrust. The complex movement of the wing in straight flapping flight generally resembles the motion of a bird's wing. The downstroke provides the main propulsion, pushing the body forward and up. During the upstroke, the wings move backward, resulting in a forward and downward force. During this phase of the wing cycle, the leading edge of the wing is turned up to increase lift and reduce the size of the downward force. The power that moves the wing through the downstroke comes principally from the chest muscles. Unlike birds, bats lack a well-developed keel on the breast bone. This gives bats a narrow chest profile that enables them to squeeze through the small openings of their roosting sites.

Flight speed is highly variable among bats. British Columbian bats fly at speeds of 7 to 36 kilometres per hour, beating their wings 10 to 20 times per second depending on their size and speed. Some species, such as the Pallid Bat (*Antrozous pallidus*), interrupt flapping flight with short bouts of gliding. Bats turn by extending and tilting the broad side of one wing in the direction of the turn while the other wing flaps normally. Bats that can turn sharply in a small space are called agile—they are usually slow-fliers, characterized by short wings. Fast-flying bats have long, narrow wings and are generally not as agile.

Flight is an expensive mode of locomotion: flying bats consume energy three to five times as fast as land-dwelling mammals of the same size. But flying requires about one-fourth the energy it would take to move the same distance on the ground. Therefore, although bats use energy quickly while flying, they can cover a large distance for a relatively low cost.

Because a bat does not walk on its hind legs, they are relatively small and weak. They show peculiar features, however, that are specializations for hanging upside down by the feet, the usual resting position for a bat. The knee joints point backward rather than forward and the feet face forward. Each foot has five toes equipped with sharp claws for gripping the surface that bats hang from. A long, cartilaginous spur called the calcar (Figure 2) is attached to one of the ankle bones and it extends along the outer edge of the tail membrane. Unique to bats, the calcar provides support for the tail membrane. The presence or absence of a distinct keel on the calcar can help to identify some species. The small, rather delicate tail extends to the end of the tail membrane in vespertilionid bats.

The head of the bat is dominated by its ears. Two British Columbian species, Townsend's Big-eared Bat (*Plecotus townsendii*) and the Spotted Bat (*Euderma maculatum*), have enormous ears—nearly half the body length. Projecting from the inner base of the ear is an erect fleshy structure called the earlet or tragus. Because its size and shape varies among species, it provides another useful trait for bat identification. See the Echolocation section for a discussion on the possible function of the tragus. The eyes are small, often hidden in the fur. This may have led to the misconception that bats are blind. All vespertilionid bats have well-developed teeth designed for chewing the hard bodies or exoskeletons of invertebrates. There are four basic types of teeth: incisors (front teeth), canines (eye teeth), premolars and molars (cheekteeth). The number of incisors and premolars varies among British Columbian bats—this is another useful characteristic in identification. The size and structure of the teeth is closely linked to diet: species that prey on large, hard-shelled invertebrates are usually equipped with the most robust teeth.

All vespertilionid bats are covered with fur. The fur is somewhat uniform in length and is not differentiated into long guard hairs and fine underfur as is found on many mammals. Bats that roost in trees have the thickest fur to protect them from open situations where they are exposed to considerable temperature variation. Although most bats are dull brown or grey, a few of our species demonstrate some beautiful colour markings. For example, the Western Red Bat (*Lasiurus blossevillii*) is rusty-red, the Spotted Bat (*Euderma maculatum*) is black with three white spots on its back, and the Hoary Bat (*Lasiurus cinereus*) and Silver-haired Bat (*Lasionycteris noctivagans*) have silver-tipped hairs that give their fur a striking frosted appearance (see the colour illustrations on the inside of the front and back covers). Colour may vary in species that are widespread across the province. Populations associated with humid coastal areas tend to be darker than those that inhabit the arid environments of the dry interior. The conventional explanation for this is that mammalian fur colour generally matches the background colour of the environment to provide camouflage. Mammals living in humid coastal areas will tend to be darker in order to blend in with the dark, dense vegetation. But this explanation is not totally convincing for bats because they are active at night and spend the daytime hidden in dark roosting sites. The fur colour of tree bats appears to be an adaptation for camouflage: the Western Red Bat looks remarkably similar to a dead leaf and the Hoary Bat resembles lichen-covered bark.

Selected References: Norberg (1987), Norberg and Rayner (1987), Vaughan (1970).

Echolocation

Another intriguing characteristic of bats is their ability to orient themselves in total darkness. Bats flying in the darkness of night can not only

Figure 3. This schematic drawing shows how a bat finds its prey using echolocation: it emits a sound pulse and listens for the echo.

avoid obstacles, but detect and capture small aerial insects. As early as 1790, scientists speculated that bats had a mysterious "sixth sense" that enabled them to see with their ears. This sixth sense is actually sonar or echolocation: the bat produces pulses of high frequency, ultrasonic sound and listens for the returning echoes (Figure 3) to obtain information on its surroundings. Vespertilionid bats produce sound pulses in the voice-box or larynx and emit them from the mouth. When the sounds hit an object they are reflected back to the bat as an echo. Results from laboratory studies suggest that these returning echoes provide the bat with a sonic image of its environment with incredibly detailed information on the size, shape and movement of objects.

The physical characteristics of the echolocation sounds are quite complex. In British Columbian species, the sound pulse generally begins at a high frequency and then rapidly drops to a lower frequency. For example, the call of the Big Brown Bat (*Eptesicus fuscus*) begins at about 80 kilohertz (a kilohertz equals 1000 cycles per second) and then quickly falls to 40 kilohertz (Figure 4). In other species, such as the Hoary Bat, the call initially decreases in freqency but is followed by a lengthy component at a constant frequency. Echolocation calls may be further complicated by the addition of harmonics or overtones.

The majority of our bats produce sounds that are above the range of human hearing. The best human ear can detect sounds with frequencies

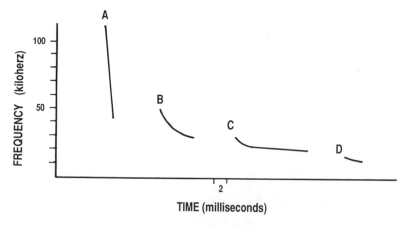

Figure 4. These sonograms (sound pictures) show the echolocation calls of four bat species: (A) Western Long-eared Myotis, (B) Big Brown Bat, (C) Hoary Bat and (D) Spotted Bat.

up to 20 kilohertz. Bat echolocation calls generally range from 20 to 120 kilohertz. But some bats produce calls that can be heard by the unaided human ear. The echolocation calls of the Spotted Bat, for example, are between 6 and 16 kilohertz, well within our range of hearing.

In general, the structure and frequency of bat echolocation calls are specific to each species. Nevertheless, in some species the echolocation calls can be highly variable: calls differ geographically among populations and even among individuals from the same local population. The reason for this variation is not well understood, but it probably results from genetic and learned differences. The design of echolocation calls may be closely related to a species' hunting strategy and the type of habitat it occupies. Bats that track their prey at long distance in open areas tend to have calls of lower freqencies. Such calls are not effective for detecting small targets and these bats tend to feed on large insects. Other species produce higher frequency calls that are most effective at short range. Their calls may sweep through a wide spectrum of frequencies to detect targets of assorted sizes. Gleaning bats tend to emit low-intensity calls, apparently to avoid confusing background echoes in a cluttered environment.

An individual bat can vary its echolocation call depending on its situation (Figure 5). When searching for insects or flying to a roost along a familiar route it emits only a few calls each second. But when it is closing in on a flying insect or approaching an unknown obstacle, it makes hundreds of calls per second. Besides changing the length and duration of calls, bats can also modify the frequency. The Hoary Bat produces constant-frequency calls when searching for prey, but when it is pursuing a target, it emits calls that sweep quickly through a range of frequencies.

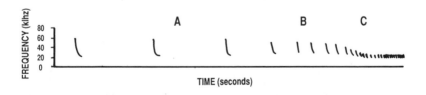

Figure 5. This sonogram illustrates the changes in a bat's echolocation calls as it (A) searches, (B) pursues and (C) captures an insect.

With each echolocation pulse generated there is a returning echo that must be detected and interpreted. Obviously this requires a sophisticated receiving system. A bat's large ears act as a funnel to pick up incoming echoes. The function of the tragus is still controversial, but some biologists think it may assist in vertical location of targets.

Despite the sophisticated echolocation systems of bats, the confrontations between bats and insects are not all one-sided. In fact, encounters between bats and some insects resemble the tactics of high-tech warfare between modern fighter planes. Some species of moths, lacewings, praying mantises and crickets can detect ultrasonic calls at distances of up to 40 metres. When a bat call is detected, the insect takes evasive action by flying in zig-zags or dropping to the ground to avoid capture. A few insects, such as some species of tiger moths, can not only detect the bat's calls but produce their own ultrasonic sounds as a defence mechanism. How the insects' sounds affect bats is not clear; they might interfere with the bat's echolocation system or perhaps they simply startle the bat. Another idea is that the sounds are a warning signal to experienced bats who have learned that this insect tastes bad.

In most regions there appears to be a relationship between the sound frequencies that moths can detect and the frequencies produced by echolocating bats. In the Okanagan region, for example, James Fullard found that moths were most sensitive to sounds that range from 30 to 75 kilohertz. Significantly, most bats in the Okanagan emit calls that range from 20 to 50 kilohertz. Bats can counter the tactics of insects by producing sounds of low intensity, short duration or frequencies outside the range audible to insects. Several British Columbian species employ these kinds of strategies. The low-frequency calls of the Spotted Bat are below the hearing range of many moths in the Okanagan. The short, quiet, high-frequency calls used by the Western Long-eared Myotis (*Myotis evotis*), a bat that feeds predominately on moths, may be inaudible to many moth species until it is too late for the moth to evade capture.

Research on the ultrasonic calls of bats has focused on their use in orientation and hunting for food. However, they may also play a role in communication. Bats may listen in on the calls of other bats to locate concentrations of insects or potential roosting sites, or to avoid neighbours in feeding areas.

Besides their ultrasonic calls, bats make an assortment of low frequency clicks, squeals and chirps that are audible to the human ear. They are used in behavioural interactions such as communication between a mother and her infant, squabbles in roosts and mate selection.

Despite the importance of sound, bats employ other senses to orient themselves and find food. Contrary to popular myth there are no known blind bats; in fact, a bat's eyesight may be more important than we once thought. The echolocation calls of most bats do not travel more than 10 metres and none reach beyond 40 metres, so bats probably rely extensively on vision for long distance movements and migration. The sense of smell is also well developed, but its role in bat behaviour is not well understood.

Selected References: Fenton (1982), Fenton (1986), Fullard et al. (1983), Fullard (1987), Griffin (1958), Roeder (1967).

Food Habits

Public curiosity about a bat's diet inspires some of the most frequently asked questions about bats. In fruit-growing regions of British Columbia a commonly asked question is: "Where are the fruit bats?" Fruit eating bats can only survive in regions where there is a year-round supply of fruit, so they are restricted to the tropics. All vespertilionid bats found in Canada feed exclusively on arthropods, an immense group of invertebrates that includes arachnids (spiders, scorpions, harvestmen), centipedes, millipedes and insects. However, nocturnal flying insects are by far the most important prey and the variety of insects taken is impressive, including moths, flies, caddisflies, midges, beetles, grasshoppers, ants and termites.

To obtain the energy, minerals and vitamins needed to fuel flight, bats must consume large quantities of food. Nursing females of temperate insectivorous species probably consume their own body weight in food each night during the summer. This rate of consumption translates into large numbers of insects, hence the importance of bats as control agents. In an experiment done in the 1960s, Little Brown Bats (*Myotis lucifugus*) released in a room of flying mosquitoes ate as many as 600 per hour. It has been estimated that the 20 million or so Mexican Free-tailed Bats (*Tadarida brasiliensis*) that inhabit Bracken cave in Texas consume about 100,000 kilograms (150 Imperial tons) of insects nightly. Worldwide, bats are without question the most important predator of nocturnal flying insects.

Insectivorous bats employ a number of strategies to capture their prey. They may hunt from continuous flight or hang from a perch and wait for a passing insect to fly or walk within range. Some species use one of these strategies, whereas others can use both depending on the availability of prey and the time of night. Most British Columbian bats hunt while flying. Although this strategy probably ensures a high encounter rate with prey items, it also requires considerable energy. The high energy requirement has resulted in bats evolving methods of capturing and consuming large numbers of insects quickly. Many species are remarkable in their ability to detect and exploit rich, ephemeral patches of food, such as a hatch of winged ants or termites. In addition, bats chew their food rapidly, up to seven times per second. The Big Brown Bat has the ability to attack an insect every three seconds. This is an amazing feat when one considers that this period includes the time required for chewing captured prey and searching for the next target.

Although the vast majority of insect-eating bats use echolocation to detect airborne prey, some hunt non-flying insects moving on vegetation or the ground by listening for their sounds. Robert Barclay and Paul Faure of the University of Calgary recently discovered that the Western Long-eared Myotis homes in on the noises made by non-flying moths. This strategy allows this bat to survive at high elevations where there are relatively few flying insects because of the cool nighttime temperatures. The Pallid Bat of the Okanagan relies on echolocation for orientation and tracking airborne prey, but it can also locate invertebrates on the ground by listening for the sounds of movement they make.

Biologists have attempted to classify bats by the type of insects they eat and labels such as "moth specialist" or "beetle specialist" can be found in many publications. However, a growing body of research indicates that bats are catholic in their choice of insect types. Not only does prey choice vary between species but also among geographical areas for the same species and even on different nights for the same local population. For example, the Big Brown Bat, found across much of North America, has been called a beetle specialist, yet in the Okanagan Valley it appears to eat mainly soft-bodied caddisflies. Obviously food habits are specific for a local area and applying information from other regions

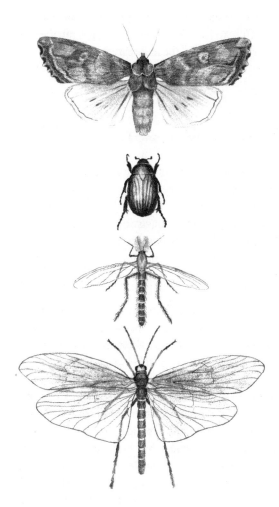

Figure 6. The four major insect groups eaten by bats in the Okanagan Valley of British Columbia (from top to bottom): noctuid moth (x1.5), beetle (x1.5), midge (x5) and caddisfly (actual size).

should be done with caution. The only information on diets for British Columbian populations is based on a few species that have been studied in the Okanagan Valley. In that region four major groups of insects—moths, beetles, midges and caddisflies—dominate the diet (Figure 6). The high incidence of aquatic insects, such as caddisflies, in the diet can be attributed to the abundance of these insects in the Okanagan on summer nights. To what extent dietary data from the Okanagan can be applied to bat populations living in the other arid grassland habitats of British Columbia is unknown. Even less is known about the food habits of bats associated with forested habitats in the province such as the coastal old-growth forests. Research on diets and hunting strategies are essential to determine the role of bats in these forested ecosytems.

Selected references: Fenton (1982), Ross (1967), Whitaker et al. (1977), Whitaker et al. (1981).

Reproduction, Development of Young and Longevity

Small mammals tend to reproduce prolifically: females produce a large number of young each year and usually breed the same summer as their birth. Vespertilionid bats do not follow this trend. All Canadian species produce only one litter each year, usually consisting of a single young. Among British Columbian species, tree bats such as the Western Red Bat and Hoary Bat are exceptions, typically producing litters of two to four young. Also, individuals in many species do not breed until their second summer.

Temperate bats mate during late summer or autumn, just before hibernation. For many species, the sexes roost separately and the period just before hibernation is the only time of year when males and females come together. In migratory species such as the Hoary Bat, mating is thought to take place on the wing during migration.

Among British Columbian species, the mating system of the Little Brown Myotis is best known. Males in this species are promiscuous and mate with as many females as possible; and females may mate with several males. There appears to be little courtship behaviour. Males initiate mating by mounting and restraining the female. Females often struggle during the early stages of copulation and males produce audible calls (5 to 12 kilohertz) that are thought to pacify females. Most mating occurs before the hibernation period when bats are still active. However, some mating occurs late in the season when the occasional active male copulates with torpid individuals of either sex.

Temperate bats delay fertilization to ensure that their young are born at the appropriate time the following spring. Fertilization does not occur immediately after mating, which is the usual pattern in mammals. Instead, the sperm cells are stored in the uterus of the female bat over winter. The following spring, after arousal from hibernation, females leave the hibernaculum and fly to the summer roost. Ovulation and fertilization take place after the females have left the hibernaculum. Most species form maternity colonies with other females of the same species. The gestation period is seven to ten weeks, depending on environmental conditions, with cold temperatures prolonging foetal development.

In Canada, baby bats are usually born in June or July. Birth dates can vary for a species because of differences in local climate, and among females in the same colony because of age. Newborns are large, often about 25% of the mother's weight, but they are underdeveloped and helpless. The skin is pink with no fur, the eyes are closed and the ears are limp. Growth is rapid, however, and within a few days the eyes open, the ears become erect, the skin develops pigmentation and the fur begins to grow. Juveniles tend to have darker, duller and shorter fur than adults. Juvenile fur is replaced by adult fur in its second month. Young bats are equipped with sharp, recurved milk teeth that are used to hang on to the mother's nipple. The milk teeth are replaced by permanent dentition about the time when the young first becomes capable of flying and eating solid food. Most of our bats attain adult size and the ability to fly in three to six weeks. Weaning (the time when young begin to eat solid food) is closely linked to the development of echolocation and the ability to fly. The age when sexual maturity is attained is known for only a few species of our bats. In the Little Brown Myotis and Townsend's Big-eared Bat, males breed in their second summer and females may be capable of breeding their first summer although they often delay breeding until the second year.

In vespertilionid bats, parental care is the sole responsibility of the female; males are usually not even present in the maternity colony. A strong bond develops between a mother and her young. Both can communicate with an assortment of calls. During the day, the young remains attached to its mother and suckles frequently; at night it is left behind at the roost while the mother departs to forage for insects. There is limited evidence that in the maternity colonies of some species, such as the Little Brown Myotis and the Fringed Myotis (*Myotis thysanodes*), a few adult females will remain behind as baby sitters. One of the more intriquing aspects of mother-infant relations is how a mother can recognize her own young, even in large colonies that may contain hundreds or thousands of baby bats. Evidently, a mother identifies her baby from its position in the colony, and its calls and its odour. There is some evidence that the infants in a few species, such as the Little Brown Myotis, can recognize the echolocation calls of their mothers. If disturbed, most female bats can fly to a new roost site carrying their young.

The mortality rate of bats is probably greatest for juveniles during their first winter. This is not surprising when one considers that in the first three months of life bats must master flight and echolocation to the extent that they can detect obstacles and capture prey. Moreover, they must consume enough food to accumulate sufficient fat reserves to survive the winter hibernation period and then find an appropriate site for hibernation. Despite the high mortality rate among juveniles, vespertilionid bats living in temperate regions can have remarkably long life spans. In the wild, the Big Brown Bat has been recorded to live 19 years, and there are records of Little Brown Myotis living more than 30 years. In Canada, relatively few animals regularly prey on bats. There are, however, records of different predators eating bats opportunistically, including snakes, hawks, owls, cats, Raccoons, skunks, weasels and Martens. Still, human activity has the most detrimental impact on bat populations.

Selected references: Grindal et al. (1992), McCracken and Gustin (1987), Racey (1982), Thomas et al. (1979), Tuttle and Stevenson (1982).

Torpor and Hibernation

Because bats are small and relatively poorly insulated, maintaining their normal body temperature of about 40°C requires a lot of energy. When food is sufficient, bats probably have little difficulty meeting their energy demands. But, when food is scarce, as it often is on a night of inclement weather or during winter, fuel consumption must be reduced.

To conserve energy during unfavourable conditions, many bats living in temperate regions have evolved the ability to lower their body temperature and metabolic rate. Over short periods of time this is called torpor; over prolonged periods it is called hibernation. The Little Brown Myotis has a resting heart rate of 100 to 200 beats per minute, and when flying, its heart rate increases to more than 1000 beats per minute; but when torpid or hibernating its heartrate is reduced to only 5 beats per minute.

When torpid or hibernating, a bat will allow its body temperature to fall below 5°C. It can raise its temperature to normal—approximately

40°C—in about 30 minutes without an external heat source. Most mammals, including bats, have two kinds of fat: white fat (the kind we carry too much of on our hips and stomachs) is well supplied with blood vessels and serves as an energy reservoir and insulator; brown fat is also an energy store, and it has a high capacity for producing heat. Brown fat is the fuel that bats use to raise their body temperature when arousing from torpor or hibernation.

Male and female bats use daily torpor differently. Unless food is extremely scarce, breeding females resist entering torpor. A high body temperature is essential for fast foetal growth and high milk production. This is why females select warm roosting sites for their maternity colonies. It is suspected that male bats, because they do not share the burden of reproduction, frequently enter daily torpor to conserve energy and minimize the time required for feeding. This is consistent with the tendency of male bats to roost separately from breeding females in cooler sites. Females of many species do not occupy high-elevation habitats where temperatures are probably too cool for them to bear young successfully.

Insect-eating bats have two choices when faced with winter food shortages: they can migrate to warmer climates where insects are available year round, or they can hibernate and subsist on fat reserves. The majority of British Columbian species rely on hibernation to cope with winter (Figure 7). Some of the migratory tree bats, such as the Silver-haired Bat and the Hoary Bat, seem to use a combination of both strategies: they migrate to a different winter range and hibernate there until spring.

Hibernation is an extension of daily torpor, but it takes much more preparation. Bats arrive at the hibernation site in late summer or autumn (September to November). Before entering hibernation they will have accumulated as much as 40% of their summer weight in fat to use as energy through the winter. Unlike other hibernating mammals, such as ground squirrels, bats cannot store food as a secondary source of energy for winter. If their body fat reserves are depleted they face starvation. Consequently, hibernation is a critical time in the life of a bat; it is a time when these mammals are extremely vulnerable to disturbance. By spring, a bat will lose 20-40% of its prehibernation weight, its fat reserves severely depleted.

Hibernacula appear to be selected on the basis of temperature and humidity, but precisely how bats make their selection is unknown.

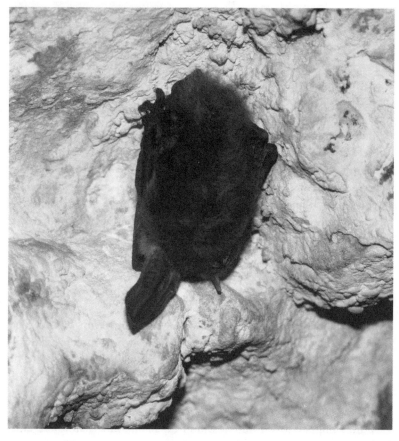

Figure 7. A hibernating Townsend's Big-eared Bat found in a small limestone cave near Williams Lake on 13 February 1990. The temperature in the cave was -7°C. (D. Nagorsen)

Each species appears to have its own prefered conditions—cool enough to allow them to maintain low metabolism, yet warm enough to keep from freezing or from burning large amounts of energy to avoid freezing. Recent research by Donald Thomas and his students suggests that humidity may be even more important to hibernating bats than temperature. Natural arousals appear to be caused by loss of water; bats occupying humid sites hibernate for longer periods before arousing to drink. Arousal is by far the most costly aspect of hibernation because it requires the expenditure of large quantities of stored fat to raise the body temperature and remain warm. Therefore, the best hibernation sites are in humid caves or mines that remain at relatively constant temperatures (0-5°C) throughout winter. Still, there are some species, such as the Big Brown Bat, that can hibernate in relatively exposed situations in buildings where there is considerable fluctuation in temperature. Many bats group in tight clusters while hibernating. It has been generally assumed that this is to prevent the loss of heat, but recent research suggests that the most important function of clustering is to reduce moisture loss.

We know that bats in the interior of British Columbia hibernate most of the winter, but little is known about the winter biology of bats in southern coastal regions where winters are mild. There is evidence that some species inhabiting these regions are periodically active in winter. For example, the California Myotis (*Myotis californicus*) has been observed active in the Vancouver region in January.

Selected References: Kunz (1987), Lyman (1970), Nagorsen et al. (in press), Ransome (1990), Thomas et al. (1990).

Roosting Sites

Roosts provide shelter from predators and inclement weather; some provide the environmental conditions appropriate for raising young. The availability of roosting sites is one of the most important factors that determines the distribution and abundance of bats. Although vespertilionid bats are best known for their use of caves, they have been found roosting in almost every conceivable type of structure: trees (hollows, under bark, in foliage), rock crevices, animal burrows, storm sewers, abandoned mines, buildings and holes under rocks. Roosting sites can be classified into three general types on the basis of when they are used: day roosts, night roosts and hibernacula. Day and night roosts are occupied by active bats in the summer; hibernacula are used for hibernation during the cold months.

Most British Columbian species spend summer days in hollows, tree cavities, buildings, rock crevices or the foliage of trees and shrubs. Hollows and buildings are used primarily by adult females congregating in large maternity colonies (Figure 8). The largest maternity colonies are in the attics of buildings. One of the largest bat colonies in the province is a maternity colony of the Yuma Myotis (*Myotis yumanensis*) in an old

Figure 8. A large cavity in a dead Ponderosa Pine is used as a maternity colony by some 70 Big Brown Bats. Tree cavities are common roosting sites in the southern Okanagan Valley. (R.M. Brigham)

church near Squilax in the interior of British Columbia. This colony contains 1500 to 2000 adult females with their young (Figures 9 and 10). Temperatures inside summer roosts can be extremely high, reaching 40°C, a characteristic that promotes the rapid growth of babies. Maternity colonies have also been found in caves or rock crevices heated by natural hot springs. The hot, humid environment created by the hot springs provides ideal conditions for developing bats (Figure 11).

Foliage-roosting species, such as the Hoary Bat and the Western Red Bat, roost alone or in small family groups comprised of a female and her young. These species demonstrate several adaptations for roosting in exposed sites. They tend to be coloured to blend in with their background and have a heavy covering of fur that insulates them from cool temperatures. The Spotted Bat and the Pallid Bat roost in narrow crevices of sheer rock cliffs (Figure 12). Because these roosts are inaccessible to humans, their environmental conditions have not been studied. Biologists are still unsure of the whereabouts of the males of most species in summer. Males turn up rarely in maternity colonies; they probably spend the summer roosting alone or in small groups in situations that are cooler than the sites selected by females.

Many insectivorous bats also use one or more night roosts, separate from the day roost, to rest and digest food between feeding flights. Night roosts occur in a variety of natural or man-made sites; they are usually more exposed and have less stable temperatures than day roosts (Figure 13). Little is known about a bat's criteria for selecting a night roost, although important factors may be environmental conditions and minimizing the commuting distance between resting and feeding areas.

Figure 9. An old church at Squilax is home to thousands of Yuma Myotis. (D. Nagorsen)

Another intriguing idea is that night roosts may be locations where bats can gather and share information about good feeding sites.

In winter, males and females of a species will share the same hibernacula in caves, mines, storm sewers, buildings, tree cavities, tree bark and rock crevices. Studies of banded bats in eastern North America have shown that some species will move several hundred kilometres between their summer roosts and winter hibernacula. Moreover, unless disturbed, bats show a strong degree of loyalty to a roosting site: they will use the same summer and winter roosts for many years.

Although British Columbia supports a diverse bat fauna in summer, only a few species have been found here in winter. One, the Silver-haired Bat, is a species that is dependent on trees—hibernating individuals have been found under the bark of Western Red-cedar and Douglas-fir. Most winter records of the Big Brown Bat are from buildings, which indicates a tolerance for cold and unstable conditions. Of the 16 species that inhabit British Columbia in summer, only the Little Brown Myotis, Western Smallfooted Myotis (*Myotis ciliolabrum*), Big Brown Bat and Townsend's Big-eared Bat have been found hibernating in caves and mine adits (Figure 14).

Not only have few species been documented to overwinter in the province, but the sizes of these winter populations are remarkably small compared to hibernation colonies found in other regions. In eastern Canada, aggregations of 10,000 to 15,000 bats hibernate in abandoned mines; in British Columbia most winter records consist of individual bats or of populations no larger than a few dozen. The largest known winter population in the province comprises about 50 bats. So where do

Figure 10. In the attic of the church at Squilax, female Yuma Myotis and their young congregate in tight clusters. (D. Nagorsen)

most British Columbian bats spend the winter? Perhaps most migrate to hibernacula in the United States, or maybe large hibernacula exist here but remain undiscovered because they are inaccessible to humans. A more likely explanation is that hibernating bats are scattered widely in small colonies throughout the province. The great number of caves, crevices, abandoned mines and trees available to bats supports this idea. An intensive inventory of potential hibernation sites in British Columbia could resolve this mystery.

The use of buildings for roosts brings bats into conflict with humans. Large numbers living in the attic of a house are a nuisance most people will not tolerate. Many colonies, however, are small and may remain undetected for years, particularly those in well insulated buildings. When guano accumulation, odours or noise from a colony necessitates a remedy, eviction and exclusion is only the safe, permanent solution. Eviction displaces the bats but does not kill them. The only inconvenience to the bats is finding a new roost. We strongly discourage killing. It is unnecessary and ineffective in permanently eliminating bats from a building.

Bats are protected under the British Columbia Provincial Wildlife Act and special permits are required to kill these animals. Be wary of advice from pest control companies about how to protect yourself from "hordes of rabid bats".

The first step in excluding bats from a building is to locate the places of entry. By watching at dusk you should be able to see where the bats exit. Closer inspection by day will reveal the holes or cracks they are using. Stains from body oils or droppings often provide clues to the locations of these openings. Any gap or crack wider than five millimetres is a potential access point. Once you have located the exits, seal them with screening or another light building material. Unlike rats and squirrels, bats cannot chew their way through a screen. A number of alternate entrances may be used, so keep watching at dusk to be sure that all openings are

Figure 11. The only known maternity colony of Keen's Long-eared Myotis is on Hotspring Island, Queen Charlotte Islands, in rock crevices heated by natural springs. (Merlin Tuttle)

Figure 12. Cliffs at Vaseux Lake in the Okanagan Valley: species such as the Spotted Bat, Pallid Bat and Western Small-footed Myotis use crevices and small openings in cliffs for day roosts. (R.A. Cannings)

Figure 13. An old mine adit in the southern Okanagan Valley is used as a temporary night roost by the Fringed Myotis. (R.M. Brigham)

Figure 14. The entrance to an old mine near Kamloops supports a hibernating population of several dozen Townsend's Big-eared Bats in winter. (D. Nagorsen)

closed off. Presumably, bats select a building because it offers good conditions for rearing young; they will be persistent in their efforts to gain access to their roost.

Ideally, all openings should be sealed in the winter when the bats have left the structure to hibernate elsewhere. If it is necessary to exclude bats in spring or autumn, close the openings at dusk after all the bats have left the roost to feed. Do not seal any openings from mid-June to the end of August. This is when flightless young are present and excluding the adults will result in the starvation and death of the young.

To maintain peace in the neighbourhood, we recommend that you warn your neighbours of any bat-removal activities. An experiment conducted near Ottawa revealed that excluded Big Brown Bats moved to the nearest available structure, usually a neighbouring house.

There is no evidence that ultrasonic noisemakers or chemical repellants (moth balls are a favourite) deter bats. In fact some ultrasonic devices actually attract bats. Moth balls (naphthalene) are toxic and the quantities required to force bats from a building may be dangerous to human health. Moreover, because moth balls evaporate quickly, the bats will usually return to the building shortly after treatment. Installing bright lights or increasing ventilation in an attic are also unreliable

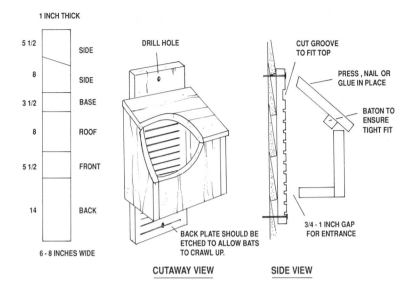

1 INCH THICK

5 1/2	SIDE
8	SIDE
3 1/2	BASE
8	ROOF
5 1/2	FRONT
14	BACK

6 - 8 INCHES WIDE

DRILL HOLE

BACK PLATE SHOULD BE ETCHED TO ALLOW BATS TO CRAWL UP.

CUTAWAY VIEW

CUT GROOVE TO FIT TOP

PRESS , NAIL OR GLUE IN PLACE

BATON TO ENSURE TIGHT FIT

3/4 - 1 INCH GAP FOR ENTRANCE

SIDE VIEW

Figure 15. Plans for a bat house, designed by Matthew Saunders.
(G. Luxton)

for removing bats. Poisons are ineffective and usually create worse problems than they are intended to solve. Poisoning increases the chance of contact between bats and people, and weakened bats are probably more susceptible to other diseases such as rabies. Remember, any poison that kills bats in a building will almost certainly be harmful to humans. Besides being ineffective, none of these approaches offers a permanent solution. Unless the access areas used by bats are closed off, the building will likely be recolonized in the future.

It is possible to invite bats to your neighbourhood without having to share your house with them. Bat houses have been used successfully in Europe and are becoming quite popular in North America. They are inexpensive and easy to build. Plans for a bat house developed by Matthew Saunders are shown in Figure 15. Detailed instructions for making a bat house can also be obtained from Bat Conservation International (see the end of the Conservation Section for the address).

Factors critical for the success of a bat house are its size and shape, the roughness of the roosting surfaces, and the distance from food and water. All inner surfaces of a bat house should be rough or horizontally grooved so that bats can get a foothold. Avoid chemically treated lumber—many wood preservatives are toxic to bats. Also do not paint, varnish or stain the house because the odours from these coatings will discourage bats. Mount the house three to six metres above the ground on a south-facing surface in a place that offers some protection from the wind. In regions with very hot climates some shade may be required. Never place a bat house where the approach is obstructed by tree limbs or power lines. Finally, be patient, most bat houses are not occupied for at least a year and sometimes only after two or three years. Some houses may never be used. The likes and dislikes of bats are not known precisely and some trial and error is necessary.

Selected References: Brigham and Fenton (1987), Kunz (1982), Nagorsen et al. (in press), Ransome (1990), Tuttle (1988).

Public Health

Folklore generally portrays a dark picture of bats, undoubtedly fostered by their close association with the darkness of night. In most western societies, bats are regarded as blood-sucking, parasite-ridden, disease-carrying vermin—this image is unfair and undeserved. Actually very few diseases are transmitted to humans by bats. With some simple precautions, bats in Canada present no risk to human health. Like most wild mammals and domestic pets, bats harbour a variety of ectoparasites including ticks, mites and fleas. Most of the parasites associated with bats however, are specific to bats and they rarely parasitize humans. Ectoparasites of temperate vespertilionid bats are not known to transmit any diseases to humans.

Histoplasmosis and rabies are two diseases associated with bats that the public should be aware of. Histoplasmosis is caused by a soil dwelling fungus that is found worldwide, especially in warm, humid regions. The growth of the fungus is enhanced by organic faeces from birds (usually pigeons or poultry) and bats. Humid caves appear to be the ideal situation for this fungus. Humans are infected by inhaling the fungal spores when they disturb dry faecal deposits. In North America, histoplasmosis is restricted mainly to the Mississippi and Ohio River valleys and adjacent areas of the United States where the warm humid climate favours the growth of this fungus. Histoplasmosis appears to be extremely rare in the Pacific Northwest and the risk of encountering this disease from bats in British Columbia is minimal. The best precaution for persons entering a bat roost that may be infected with histoplasmosis is to wear a respirator that can filter particles as small as two microns in diameter.

By far the greatest fear we humans have of bats is that they can transmit rabies, a fatal viral disease. Rabies can infect virtually any mammal, but carnivores and bats seem to be most susceptible. One of the most frightening symptoms is paralysis, especially of the hind limbs and throat muscles. Paralysis of the animal's throat muscles prevents it from swallowing, and the accumulation of saliva gives it the appearance of frothing at the mouth. The virus is usually transmitted in the saliva by a bite. Once in the host, the rabies virus progresses along nerves to the spinal cord and brain, eventually invading all nervous tissue and the salivary glands.

The incubation period (time from exposure to development of symptoms) is usually several weeks, although in a few cases it may be up to a year. Some mammals may develop an excitable or furious form of rabies in which they become extremely aggressive. North American bats rarely develop furious rabies and they may appear normal except for a gradual weakness and loss of flying abilities that results from the developing paralysis. Outbreaks of rabies among bat populations have never been observed and even in large colonies containing thousands of bats infections seem to involve only the occasional individual.

Rabies was not discovered in North American bats until 1953. The number of infected bats detected is usually proportional to the intensity of the testing effort and method of sampling bats. In North America, 5-10% of all bats submitted for rabies testing prove to be infected. This may appear to be a high incidence, but this represents an extremely biased sample because the bats submitted for testing are ill, dead or behaving abnormally. Of bats captured in normal foraging flight in the wild, only a small portion (0.1-0.5%) were infected. In North America, about one in a thousand clinically-normal bats carry the disease, about the same level as found in pigs. To put the risk of rabies from bats in perspective, only about ten human deaths are known to be caused from bat rabies in the United States and Canada. Rabies is a relatively rare disease in British Columbia. Unlike the prairie provinces and Ontario, where there is a high incidence of rabies in Red Foxes and Striped

Skunks, the disease is virtually unknown in British Columbian carnivores. Nevertheless, a few rabid bats are diagnosed each year in British Columbia.

Most human exposures to bat rabies come from bites received while handling bats. For people at risk—e.g., biologists studying bats or veterinarians—a vaccine is available that provides protection from the disease. For most individuals, however, this is unnecessary and the best precaution is to avoid handling bats. Be particularly wary of bats laying on the ground or behaving abnormally. Children should be cautioned about the dangers of touching bats or, for that matter, any unfamiliar wild mammal. Pets, especially cats, have been known to bring disabled bats into the home. We recommend vaccination of dogs and cats.

If it is necessary to handle bats then there are a few precautions you can take. Wear leather gloves when handling live bats; wear disposable rubber gloves to pick up dead bats, or you can improvise by covering your hand with a plastic bag. If you are concerned about possible exposure to a rabid bat, contact your local health unit or district veterinarian. Development of rabies can usually be arrested if treatment begins with a vaccine immediately after exposure.

Selected references: Constantine (1988), Tuttle (1988).

Conservation

Eight of the province's bats, half our native species, are currently listed on the Ministry of Environment's Red and Blue lists of potentially endangered and threatened species. For most species, the biology and population status in British Columbia is poorly known. With no long-term data on population numbers, we have no clear indication if numbers are stable or declining. There is evidence of population declines for many vespertilionid species over the past few decades in Europe and North America. Two bats in the United States appear on the Federal Endangered Species List and several others have undergone serious declines. All fifteen bats native to Great Britain are now considered vulnerable or endangered. Major threats to bat populations everywhere are extermination, disturbance of roosting sites, environmental contaminants (such as pesticides), and loss of habitat.

Extermination of roosting bats is one of the most serious threats.

Large maternity colonies in buildings are considered a nuisance and are often destroyed with little concern for the impact on bat populations. As we explained in the Roosting Sites section, one can usually evict bats without resorting to killing them. It is often argued that bats associated with buildings are the common species, but we know so little about population numbers that it is difficult to know which species are common. Moreover, some bats that roost in buildings may be on lists of threatened or endangered species. For example, the only known maternity colony in British Columbia for the Fringed Myotis, a species on the provincial Blue List, was found in the attic of a house near Vernon.

Hibernating bats are especially vulnerable to vandalism or disturbance. Even if they are not directly killed or injured by human activity, disturbance from hibernation will cause a bat to use up fat reserves prematurely—later, it may die of starvation. Although large hibernating colonies have yet to be found in British Columbia, several species rely on caves or abandoned mines for hibernacula. The most susceptible to disturbance is Townsend's Big-eared Bat. Small hibernating populations occupy a number of caves and mines throughout southern British Columbia and all of these hibernacula are potentially threatened by human activity.

Because they eat insects, bats are susceptible to poisoning by pesticides, especially by organochlorines such as DDT. Although single dosages may not be lethal, these chemicals tend to be retained in the fatty tissues and bats can accumulate significant residues over their long life spans. These residues are released when bats use their fat reserves during periods of stress such as migration or hibernation. Some organochlorine pesticides may be passed on to nursing young through the mother's milk.

Bats may be exposed directly to poisonous chemicals when their roosts are treated by exterminators. But most exposure probably results from the spraying of forests and agricultural crops. A recent study in Great Britain revealed that bats also can pick up toxic levels of pesticides from contact with roofing materials and building timbers treated with preservatives such as lindane and PCP (pentachlorophenol). The effect of other environmental contaminants is not clear. High levels of mercury and lead have been found in several bat populations but the significance of these levels is unknown.

Habitat loss, especially the loss of roosting sites, is also of critical importance. In British Columbia, forestry operations probably have the greatest impact on bat habitat. Clear cutting large tracts of timber, removal of snags (standing dead trees), and the intensive management of young forests alter the forest structure and the availability of roosting sites. Tree-inhabiting species, such as the Hoary Bat, Western Red Bat and Silver-haired Bat, are adversely affected by deforestation. Keen's Long-eared Myotis (*Myotis keenii*), a rare bat restricted almost entirely to the coastal forests of British Columbia, has received a lot of publicity recently because it is assumed to be dependent on old-growth forests. There is even speculation that it is a bat version of the Spotted Owl. Unfortunately, as is true for most of our bats, little is known about the roosting requirements of this species. Even species that now roost primarily in buildings probably relied extensively on trees before the proliferation of human dwellings. For example: in the Okanagan, maternity colonies of the Big Brown Bat, a species usually associated with buildings, are situated mostly in cavities of dead Ponderosa Pines.

Wildlife biologists have given considerable attention to how large mammals and some smaller mammals such as rodents use forests of different ages, but little research has been done on their use by bats. The only comprehensive study done in the Pacific Northwest was Donald Thomas's research in unmanaged coastal forests of Washington and Oregon. Thomas observed that the activity of tree-roosting bats was highest in old growth forests and concluded that these forests with their abundant old trees and large snags offered more roosting opportunities than younger forests. Large snags may be especially important for bats because their cavities and loose slabs of bark provide many potential roosting sites. To help the forestry industry develop practices sensitive to bat populations, it is imperative that research is carried out to determine the roosting habits of our forest species and the impacts of such practices as tree thinning and snag removal.

Education to change people's negative perception of bats is obviously critical to ensure the survival of these animals. One of the most important contributions that the average person can make to bat conservation is simply being more tolerant of these mammals. For individuals especially interested, we recommend that you join Bat Conservation International, Post Office Box 162603, Austin, Texas, 78716, U.S.A. BCI is a non-profit organization that promotes bat conservation and provides educational materials (slides, books, education kits, etc.) about bats in general.

Selected References: Barclay et al. (1980), Clark (1981), Mitchell-Jones et al. (1989), Thomas (1988), Tuttle (1988), Wildlife Branch (1993).

Studying Bats

Because they are volant and nocturnal, bats are difficult to study in the wild; until recently, research on these animals has lagged behind the study of other mammals. It is important to remember that Donald Griffin's discovery of echolocation was just 50 years ago. Within the past few decades, the development of new devices for live capture, sophisticated electronic bat detectors, night vision scopes and miniature radio transmitters has greatly increased the number of tools available to biologists studying bat ecology and behaviour.

Despite the new technology, the most popular device for capturing live bats is the mist net (Figure 16), developed some 300 years ago by Japanese hunters to catch birds for food. A mist net consists of four sections or shelves that are supported by cross lines. The net is strung taut between two poles. The nets are about two metres high and can range from six to thirty-six metres in length. The net is made of fine nylon,

Figure 16. Mist nets strung across the Okanagan River. (R.M. Brigham)

terylene or monofilament material similar to hair nets. Mist nets are effective almost anywhere bats fly, but most captures are made near roosts, over small pools and streams or along trails used as flyways. Mist nets are inexpensive, light-weight, easy to use in the field and capable of covering a relatively large area. But they require constant attention: captured bats must be quickly removed or they will escape or become hopelessly entangled. And mist nets will not work for all bats; some species are renowned for their proficiency at avoiding them.

The harp trap (Figure 17) is another device for capturing live bats. It consists of two frames of aluminum tubing each about two square metres joined together 70-100 mm apart. Each frame has a bank of vertical strands of monofilament fishing line spaced at 25-mm intervals. The trap is supported by four tubular extension legs and a canvas bag partially lined with polyethylene is suspended beneath the device. It collects bats as they hit the monofilament lines and slide down. We do not know why the harp trap is so effective. Bats seem to be able to avoid the first bank of lines, but they are unable to negotiate the second bank. Temporarily trapped between the two sets of vertical lines, they drop or flutter unharmed into the bag suspended below. As with mist nets, harp traps are most productive when set in natural flyways on trails, between trees or at roost entrances. The advantage of the harp trap is that it requires minimal monitoring; captured bats cannot escape from the bag and are reasonably protected from predators and inclement weather. The researcher can set the trap at dusk and then return early the next morning. Another advantage is that species adept at avoiding mist nets can be caught in harp traps. The disadvantages of a harp trap are its high cost, relatively small area of coverage and lack of portability.

Bats can be difficult to observe at night while flying in

Figure 17. A harp trap. (R.M. Brigham)

their natural habitats. Attaching small light-tags to captured bats enables biologists to observe them roosting or foraging from a distance. Inexpensive light-tags can be made using Cyalume, a greenish-yellow phosphorus compound, mixed with a peroxide-based reactant to produce a bright, cold, non-toxic light. Cyalume sticks can be purchased in sealed tubes made of flexible plastic in which the peroxide is isolated in a glass vial. Bending the tube breaks the vial and initiates the reaction. The resulting chemiluminescent liquid is then drawn off in a syringe and placed in gelatin or glass capsule. The capsule is glued to the dorsal or ventral fur of the bat with surgical adhesive. Depending on its size, the light-tag usually glows brightly for about two hours and is visible to the naked eye for about 200 metres. By using different colours it is possible to mark males and females or different species. In the south Okanagan, light tags have been used successfully to compare hunting behaviour of the California Myotis and the Western Small-footed Myotis, as well as the habitat use and foraging behaviour of the Little Brown Myotis and Yuma Myotis.

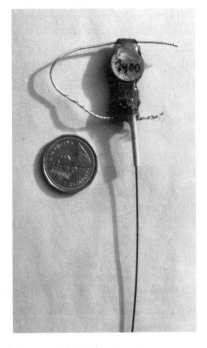

Figure 18. This small radio transmitter, weighing about two grams, could be used on large bats such as the Hoary Bat. (R.M. Brigham)

A more expensive method of tracking animals is with radiotelemetry, which involves attaching tiny, battery-powered radio-transmitters to the bat's fur. The transmitter emits a pulsing signal detectable with a portable receiver and directional antenna. As with light-tags it is important to keep the transmitter as small as possible to ensure that the bat's flying abilities are not impeded. Transmitters used on small bats weigh about 0.7 grams (Figure 18). As a rule of thumb they should not exceed

5% of the bat's body mass; therefore, bats weighing 12 to 15 grams are the smallest species that can be studied with radiotelemetry. These small transmitters can be expected to last 10 to 30 days although they usually fall off before they stop beeping. Radiotelemetry has provided many details for a number of bat species about movements between roosts, home range size, foraging behaviour and activity periods. In British Columbia, roosting sites of the Pallid Bat and Spotted Bat have been located using radiotelemetry.

Bat detectors (Broadband Ultrasonic Detectors) are sensitive to sounds of 10 to 180 kilohertz. They can be used to study bat activity, distribution and habitat preferences. For field use, small detectors such as the QMC Mini Bat Detector (QMC Instruments limited, 229 Mile End Road, London, England, E14AA, United Kingdom) are inexpensive and highly portable. The detector converts the high-frequency echolocation calls into lower-frequency sounds audible to the human ear. Because echolocation calls are often unique to a bat species, detectors often make it possible to identify some species from their calls much the same way an ornithologist can identify bird species from song. The calls of the Western Red Bat, Hoary Bat and Spotted Bat are unmistakable; but distinguishing the calls of our nine species of *Myotis* is not possible with a simple bat detector.

While biologists require some highly technical equipment required to study bats in detail, interested naturalists or lay-persons can still make a significant contribution to our knowledge of bats in British Columbia. The distribution and roosting habitats of most species in the province are poorly documented and any observational records are useful. The identification key in this book will assist the user in identifying live bats. Nevertheless, some species are difficult to identify from external features even for the expert, and we urge naturalists to document any observations with colour photographs. The date, location and roosting situation should also be recorded. Summer maternity colonies and winter hibernacula are of particular interest. These colonies are extremely sensitive to human disturbance and should only be examined by experts. Report any observations of maternity or hibernating colonies to the Royal British Columbia Museum or the Ministry of Environment. Bats found dead can be brought in to your local Ministry of Environment office or the Royal British Columbia Museum for identification. However, before you consider handling any bats, we urge you to

read carefully the precautions for rabies (see the Public Health section).
Selected References: Kunz (1988), Thomas and West (1989).

Bats in British Columbia

Encompassing some 950,000 square kilometres and spanning 11 degrees of latitude and 25 degrees of longitude, British Columbia has the most diverse physiography and climate of any Canadian province. Rows of north-south oriented mountain ranges dominate the landscape (Figure 1). They play a major role in the province's climate by intercepting Pacific weather systems as they move eastward and creating alternating wet-dry belts. The wettest regions are along the Pacific coast, especially the western slopes of the coastal mountain ranges and the west coasts of Vancouver Island and the Queen Charlotte Islands. East of the Coast Mountains, rain shadows create a large arid region: the interior plateau. Most extreme arid conditions are found in some of the southern interior valleys such as the Okanagan and Thompson river valleys. Other wet-dry belts are associated with the Cassiar, Rocky, Cariboo, Monashee, Selkirk and Purcell mountain ranges.

The vegetation of the province is predominately coniferous forest, although deciduous forest is associated with northern boreal regions and riparian habitats along rivers and lakes. Grassland and shrub-steppe habitats occur in some of the arid, southern interior valleys. Grassy alpine tundra and scrubby willow-birch habitats are common in northern British Columbia and at high elevations in southern parts of the province.

A number of ecological classifications have been developed for British Columbia. Cowan and Guiguet, in the original *The Mammals of British Columbia*, proposed 13 biotic areas for the province. Two more recent classifications systems are the system of 14 biogeoclimatic zones developed by the Ministry of Forests (see colour fold-out map) and the 10 ecoprovinces developed by the Ministry of Environment (Figure 19). The biogeoclimatic zones define areas that have relatively homogeneous climates and characteristic vegetations. In mountainous terrain the zones are related to elevation. An ecoprovince is a broad geographical area with a consistent climate and terrain.

British Columbia supports the highest bat diversity of any Canadian

Figure 19. The ten ecoprovinces of British Columbia. (Wildlife Branch, Ministry of Environment, Lands and Parks)

province (16 species), largely because of its environmental diversity. On a provincial scale, bat diversity is greatest at low latitudes and elevations. The arid south Okanagan Valley is the most diverse region with 14 summer resident species. In fact, the Okanagan has the greatest variety and population densities of bats of any region in Canada. Numbers of bat species decline sharply at high latitudes. Only two species are known to occur in the Atlin area of northwestern British Columbia and only one species has been documented in the Fort Nelson area in the northeast. Numbers of bats also decline with increasing elevation. Even in southern regions, few species are found at elevations above 1500 metres.

These general trends are reflected in the distribution of bats in the 14 biogeoclimatic zones (Table 1). Bats are most diverse in the Bunchgrass and Ponderosa Pine zones; the number of species is greatly reduced in zones associated with northern boreal regions or high elevations. The high bat diversity in the Bunchgrass and Ponderosa Pine zones is significant because these zones cover only a small portion of the province's total land mass. Moreover, grassland habitats in these zones, especially in southern regions of the province, are threatened from cultivation and urbanization (Figures 20, 21 and 22). The trend for decreasing bat diversity at higher latitudes is also shown in the distribution of bats in the nine terrestrial ecoprovinces (Table 2). Highest bat diversity is in the southern ecoprovinces.

Because of their mobility, bats are less impeded by physical barriers than the other terrestrial mammals. In contrast to the ranges of small

rodents and insectivores, bat distributions are not affected by major rivers such as the Fraser, Thompson and Skeena. Bats have also colonized most of the islands off the coast. In fact, bats are the most diverse group of small mammals on many British Columbian islands. The Queen Charlotte Islands have only one native shrew and one species of mouse, but support four bat species. Similarly, Vancouver Island is home to only six native species of mice and shrews but has ten bat species, the same number as the coastal mainland.

Climate and the availability of roosting sites are major factors that determine the distributional patterns and species diversity of bats in British Columbia. Roosting sites are probably scarce in scrubby subalpine and alpine habitats and this may contribute to the low diversity of bats at

Figure 20. Ten bat species are found in the coastal forests of British Columbia, including the rare Keen's Long-eared Myotis, a species that appears to be restricted to this region. (D. Nagorsen)

Table 1. Distribution of 16 bat species in the 14 biogeoclimatic zones of British Columbia.

SPECIES	Alpine Tundra	Bunchgrass	Boreal White and Black Spruce	Coastal Douglas-fir	Coastal Western Hemlock	Engelmann Spruce-Subalpine Fir	Interior Cedar-Hemlock	Interior Douglas-fir	Mountain Hemlock	Montane Spruce	Ponderosa Pine	Spruce-Willow-Birch	Sub-Boreal Pine-Spruce	Sub-Boreal Spruce
California Myotis (*Myotis californicus*)		•		•	•	?	•	•		•	•			
Western Small-footed Myotis (*Myotis ciliolabrum*)		•						•			•			
Western Long-eared Myotis (*Myotis evotis*)		•	•	•	•	•	•	•	?	•	•		•	•
Keen's Long-eared Myotis (*Myotis keenii*)				?	•									
Little Brown Myotis (*Myotis lucifugus*)	?	•	•	•	•	•	•	•	?	•	•	•	•	•
Northern Long-eared Myotis (*Myotis septentrionalis*)			•				•							
Fringed Myotis (*Myotis thysanodes*)		•						•			•			
Long-legged Myotis (*Myotis volans*)		•	•	•	•	•	•	•	?	•	•		?	•
Yuma Myotis (*Myotis yumanensis*)		•		•	•		•	•			•			
Western Red Bat (*Lasiurus blossevillii*)		•			•									
Hoary Bat (*Lasiurus cinereus*)		•		•	•	•	•	•		•	•			
Silver-haired Bat (*Lasionycteris noctivagans*)		•	•	•	•	•	•	•	•	•	•	•	•	•
Big Brown Bat (*Eptesicus fuscus*)		•	•	•	•		•	•		•	•		•	•
Spotted Bat (*Euderma maculatum*)		•						•			•			
Townsend's Big-eared Bat (*Plecotus townsendii*)		•		•	•		•	•			•			
Pallid Bat (*Antrozous pallidus*)		•									•			
TOTAL SPECIES	0?	14	6	9?	11	5?	10	12	1?	7	13	2	4?	5

Table 2. Distribution of 16 bat species in the nine terrestrial ecoprovinces of British Columbia.

SPECIES	Coast and Mountains	Georgia Depression	Southern Interior	Central Interior	Southern Interior Mountains	Sub-Boreal Interior	Northern Boreal Mountains	Boreal Plains	Taiga Plains
California Myotis (*Myotis californicus*)	•	•	•	•	•				
Western Small-footed Myotis (*Myotis ciliolabrum*)			•	•					
Western Long-eared Myotis (*Myotis evotis*)	•	•	•	•	•	•	•		
Keen's Long-eared Myotis (*Myotis keenii*)	•	•							
Little Brown Myotis (*Myotis lucifugus*)	•	•	•	•	•	•	•	•	?
Northern Long-eared Myotis (*Myotis septentrionalis*)					•			•	?
Fringed Myotis (*Myotis thysanodes*)			•	•					
Long-legged Myotis (*Myotis volans*)	•	•	•	•	•	•	•		
Yuma Myotis (*Myotis yumanensis*)	•	•	•	•	•				
Western Red Bat (*Lasiurus blossevillii*)	•		•						
Hoary Bat (*Lasiurus cinereus*)	•	•	•	•	•				
Silver-haired Bat (*Lasionycteris noctivagans*)	•	•	•	•	•	•	?	•	
Big Brown Bat (*Eptesicus fuscus*)	•	•	•	•	•	•		•	
Spotted Bat (*Euderma maculatum*)			•	•					
Townsend's Big-eared Bat (*Plecotus townsendii*)	•	•	•	•	•				
Pallid Bat (*Antrozous pallidus*)			•						
TOTAL SPECIES	11	10	14	12	10	5	3?	4	0?

Figure 21. The southern Okanagan Valley supports the most diverse and abundant bat fauna in Canada. It is home to rare species such as the Spotted Bat, Pallid Bat, Fringed Myotis and Western Red Bat. (R.A. Cannings)

higher elevations. In contrast, the Okanagan Valley has a rich assortment of cliffs, rock faces and open forests in which bats can roost. Temperature, however, seems to have the greatest impact on bat distributions in the province. Even in summer, the cool night temperatures at high latitudes or elevations may cause bats to spend many nights torpid, thereby reducing opportunities for hunting and fattening for winter. Also, flying insects are scarce in cooler regions. It is not surprising that only a few species can cope with these conditions. Most bats prefer the hot, arid valleys, such as the Okanagan. Summer night temperatures are usually warm enough for foraging, and they promote extremely high densities of flying insects, especially over lakes and rivers.

The winter range of most of our bats is also closely related to temperature and to the availability of suitable hibernacula. Bats have been found hibernating as far north as Prince George (54°N), and there is evidence to suggest that a few species may hibernate at even higher latitudes. Maternity colonies of the Little Brown Myotis have been found as far north as 60°N in British Columbia—presumably these

Figure 22. A subalpine area near the tree line in the Coast Mountains: only a few species, such as the Western Long-eared Myotis and Little Brown Myotis, are regularly found at high elevations in British Columbia. (D. Nagorsen)

populations overwinter in extreme northern regions because this species does not usually migrate more than a few hundred kilometres from its summer range. Donald Thomas's research on the Little Brown Myotis suggests that overwintering populations are limited by the relationship between the lengths of the summer growing season and the winter hibernation period.

Selected References: Humphrey (1975), Meidinger and Pojar (1991).

Taxonomy and Nomenclature

In 1758, a Swedish botanist named Carl Linnaeus invented a classification system for all living things. The Linnaean system of classification uses a hierarchy of taxonomic categories: class, order, family, genus, species. Although the definition of species is somewhat contentious, most biologists consider a species to consist of populations that are capable of interbreeding. Each species has a unique scientific name or binomen consisting of the genus and species name. By convention, the binomen is italized with the first letter of the genus capitalized and the species name all in lower case. For example, the scientific name for the Little Brown Myotis is *Myotis lucifugus*.

Closely related species that share a number of similar traits are usually grouped in the same genus. Some 90 species are recognized in the genus *Myotis*; 9 are found in British Columbia. Some species have distinct geographic races or subspecies and these are recognized formally by taxonomists with a trinomen. *Myotis lucifugus carissima*, for example, is a pale race of the Little Brown Myotis that inhabits arid areas of the western United States and the dry interior of British Columbia.

Species can be grouped into higher taxonomic categories based on their presumed relationships. The 1000 known species of bats are grouped into 17 families; only 2 of these families, Vespertilionidae and Molossidae, are represented in British Columbia. (Only one species of molossid has been recorded in the province and this was an accidental occurrence. See Species Accounts.) All bats belong to the mammalian order Chiroptera.

Taxonomy is a dynamic science and there have been important changes in the taxonomy of bats since Cowan and Guiguet's *The Mammals of British Columbia* was last revised in 1965. Townsend's Big-eared Bat is now placed in the genus *Plecotus* rather than *Corynorhinus* and the the Big Free-tailed Bat is now considered by most taxonomists to be a member of the genus *Nyctinomops* instead of *Tadarida*. More significantly, three bats have been split into separate eastern and western species. Western populations of the Small-footed Myotis (*Myotis leibii*) and Red Bat (*Lasiurus borealis*) are now considered sufficiently distinct to be recognized as full species: the Western Small-footed Myotis (*Myotis ciliolabrum*) and Western Red Bat (*Lasiurus blossevillii*). Similarly, *Myotis keenii* has been split into two species: Keen's Long-eared Myotis (*Myotis keenii*), a Pacific coast species, and the Northern Long-eared Myotis (*Myotis septentrionalis*), a species that ranges from the Rocky Mountains to eastern Canada. These taxonomic changes have important implications beyond simple name changes. For example: although there is considerable published literature on Red Bats, most is based on research done on the eastern species (*Lasiurus borealis*) and it is not known how applicable this information is to the western species (*Lasiurus blossevillii*).

Recent research has also resolved the taxonomic relationships between the Little Brown Myotis (*Myotis lucifugus*) and Yuma Myotis (*Myotis yumanensis*) in British Columbia. The two species can be extremely difficult to distinguish and in some regions a few individuals appear intermediate in their external and cranial features. Some authorities interpreted these intermediate individuals as hybrids and evidence for interbreeding between the two species. However, in a genetic study of populations from the Okanagan Valley, Robert Herd and Brock Fenton found no evidence for hybridization. All individuals, even those intermediate in cranial features, could be unequivocally identified as either *Myotis yumanensis* or *Myotis lucifugus* from their genetic characteristics.

The scientific names of bat species used in this book are based on *The Mammals of British Columbia: A Taxonomic Catalogue*. It contains detailed information on the taxonomy of British Columbian bats including citations to subspecies descriptions.

There is no universally accepted list of mammalian common names equivalent to the American Ornithologists Union checklist of bird names. Common names used in this book were taken from *The Vertebrates of British Columbia: Scientific and English Names*. One of the goals of this checklist is to establish standard common names for the province's vertebrates. The reader should be aware, however, that the common names for many bats are somewhat contrived and they are rarely used by scientists. Moreover, several species have other widely-used common names. For example, the Little Brown Myotis (*Myotis lucifugus*) is often referred to as the Little Brown Bat and Townsend's Big-eared Bat (*Plecotus townsendii*) is known as the Lump-nosed Bat or Western Big-eared Bat in some regions.

Selected References: Herd and Fenton (1983), Nagorsen (1990*a*), Nagorsen (1990*b*), van Zyll de Jong (1985).

CHECKLIST OF BRITISH COLUMBIA BATS

Families and genera are ordered according to their generally accepted phylogenetic arrangement. Species within a genus are arranged alphabetically.

Order Chiroptera: Bats

Family Vespertilionidae: Vespertilionid Bats

Myotis californicus (Audubon and Bachman)	California Myotis
Myotis ciliolabrum (Merriam)	Western Small-footed Myotis
Myotis evotis (H. Allen)	Western Long-eared Myotis
Myotis keenii (Merriam)	Keen's Long-eared Myotis
Myotis lucifugus (Le Conte)	Little Brown Myotis
Myotis septentrionalis (Trouessart)	Northern Long-eared Myotis
Myotis thysanodes Miller	Fringed Myotis
Myotis volans (H. Allen)	Long-legged Myotis
Myotis yumanensis (H. Allen)	Yuma Myotis
Lasiurus blossevillii (Lesson and Garnot)	Western Red Bat
Lasiurus cinereus (Palisot de Beauvois)	Hoary Bat
Lasionycteris noctivagans (Le Conte)	Silver-haired Bat
Eptesicus fuscus (Palisot de Beauvois)	Big Brown Bat
Euderma maculatum (J. A. Allen)	Spotted Bat
Plecotus townsendii Cooper	Townsend's Big-eared Bat
Antrozous pallidus (Le Conte)	Pallid Bat

Family Molossidae: Molossid Bats

Nyctinomops macrotis (Gray)	Big Free-tailed Bat
(Accidental occurrence.)	

IDENTIFICATION KEYS

Two separate identification keys, one based on external features and the other on cranial and dental characters, are provided to assist in the identification of the province's bats. The keys are dichotomous with the diagnostic characteristics arranged into couplets; each couplet offers the user two mutually exclusive choices (labelled a or b). To identify a bat, begin with couplet number one and select a or b. This will give you either a species name or direct you to another couplet in the key. By working through the various steps in the key you will eventually arrive at an identification. We tried to avoid subjective traits (e.g., slightly darker than or slightly larger than) and instead emphasized presence or absence of features or absolute size measurements. To simplify the key, the diagnostic criteria in each couplet are limited to one or two characteristics. Many identification keys incorporate locality information (e.g., found only in the Okanagan Valley). We deliberately avoided geography in the keys. Accurate locality information is sometimes lacking; moreover, we do not know the precise range for many bats in the province and excluding a species on the assumption that it is absent from an area could be misleading. Once you obtain an identification, consult the appropriate species account to see if it is consistent with your determination from the key.

External Features

The key to external features can be used on restrained live bats held in the hand, dead bats or museum study skins. (Read the Public Health section on rabies before you consider handling live or dead bats.) The

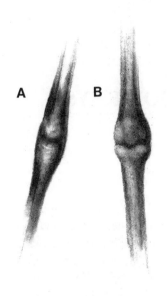

Figure 23. The finger joints of (A) an adult bat and (B) a juvenile bat.

diagnostic characteristics used are of adults. Nursing young cannot be identified with this key and some of the size criteria cannot be applied to immature bats that have not reached their full adult size. You can distinguish immatures from adults by their swollen finger joints (Figure 23) and dark fur. The only tools that you will require are a small millimetre rule to measure forearm and ear lengths (Figure 24) and a hand lens to verify the presence of minute hairs on the edge of the tail membrane of several species. The user should be aware that there are several species of bats in the province that can be extremely difficult to distinguish and their diagnostic characters must be checked carefully—e.g., Keen's Long-eared Myotis (*Myotis keenii*) vs. Western Long-eared Myotis (*Myotis evotis*) and Yuma Myotis (*Myotis yumanensis*) vs. Little Brown Myotis (*Myotis lucifugus*). Some individuals are impossible to identify with a absolute certainty by their external features alone and positive identification may require an examination of their skulls by an expert.

1a   Ear length greater than 28 mm 2
1b   Ear length less than 28 mm 4

2a   Fur on back black with three white spots on rump and shoulders
     .... Spotted Bat (*Euderma maculatum*)
2b   Fur on back not black and lacking white spots 3

3a   Nose with two prominent bumps, forearm length 40-45 mm
     .... Townsend's Big-eared Bat (*Plecotus townsendii*)
3b   Nose lacking two prominent bumps, forearm length 48-57 mm
     .... Pallid Bat (*Antrozous pallidus*)

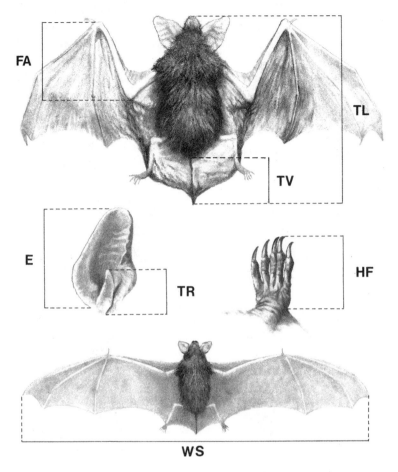

Figure 24. The standard external measurements of a bat: *TL*, total length; *TV*, tail vertebra; *HF*, hind foot; *E*, ear; *TR*, tragus; *FA*, forearm; *WS*, wingspan.

4a Fur orange or rusty red in colour
.... Western Red Bat (*Lasiurus blossevillii*)
4b Fur not orange or rusty red 5

5a Fur on back with frosted or silver-tipped hairs 6
5b Fur on back without frosted or silver-tipped hairs 7

6a   Upper surface of tail membrane covered with fur, forearm length 50-57 mm .... Hoary Bat (*Lasiurus cinereus*)

6b   Upper surface of tail membrane furred at base, forearm length 39-44 mm .... Silver-haired Bat (*Lasionycteris noctivagans*)

7a   Visible fringe of hairs on outer edge of tail membrane (Figure 25A) .... Fringed Myotis (*Myotis thysanodes*)

7b   No visible fringe of hairs on outer edge of tail membrane (Figure 25B) 8

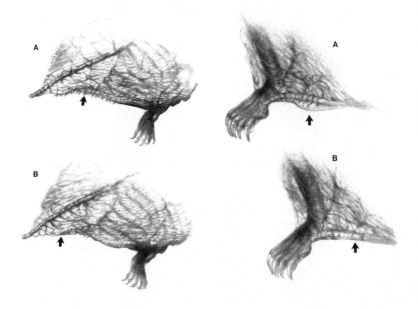

Figure 25.   Figure 26.

8a   Calcar with a prominent keel (Figure 26A) 9

8b   Calcar without a prominent keel (Figure 26B) 12

9a   Forearm length greater than 35 mm, hind foot greater than 9 mm 10

9b   Forearm length less than 35 mm, hind foot less than 9 mm .. 11

Figure 27.

10a Underwing furred outward to a line extending from knee to elbow (Figure 27A), forearm length 34-44 mm

.... Long-legged Myotis (*Myotis volans*)

10b Underwing not furred outward to a line extending from knee to elbow (Figure 27B), forearm length 43-52 mm

.... Big Brown Bat (*Eptesicus fuscus*)

11a Fur on back pale blond to orange-yellow, contrasts sharply with blackish ears, face and wings; length of bare area on snout about 1.5 times width across nostrils (Figure 28A)

.... Western Small-footed Myotis (*Myotis ciliolabrum*)

11b Fur on back chestnut to brown, not in sharp contrast to colour of ears, face and wings; length of bare area on snout about equal to width across nostrils (Figure 28B)

.... California Myotis (*Myotis californicus*)

12a Ears long (14-22 mm), extending well beyond tip of nose when pushed forward 13

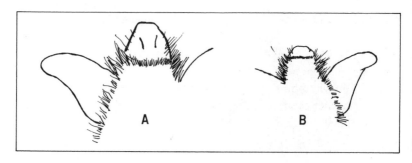

Figure 28. The bare area on the snout (A) about 1.5 times the width across the nostrils and (B) about equal to the width across the nostrils.

12b Ears short (10-16 mm), not extending well beyond tip of nose when pushed forward 15

13a Ears black, extending more than 5 mm beyond tip of nose when pushed forward ..... Western Long-eared Myotis (*Myotis evotis*)
13b Ears dark but not black, extending less than 5 mm beyond nose when pushed forward 14

14a Poorly defined dark spot on shoulders, minute hairs on edge of tail membrane (visible with hand lens)
.... Keen's Long-eared Myotis (*Myotis keenii*)
14b No dark spot on shoulders, few or no hairs on edge of tail membrane ...... Northern Long-eared Myotis (*Myotis septentrionalis*)

15a Fur on back long, sleek and glossy, forearm length usually greater than 36 mm .... Little Brown Myotis (*Myotis lucifugus*)
15b Fur on back short and dull, forearm length usually less than 36 mm .... Yuma Myotis (*Myotis yumanensis*)

Cranial and Dental Characters

This key is designed to be used on cleaned skulls. It is intended for use with museum specimens or skulls that may be found in caves or owl pellets. You will require a hand lens or binocular dissecting microscope and dial vernier calipers. Different types of teeth referred to in the key are identified in Figure 29; skull measurements are illustrated in Figure 30.

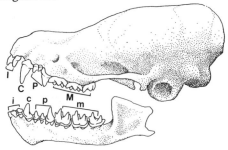

Figure 29. The four types of teeth in a bat's skull: *I*, upper incisors; *i*, lower incisors; *C*, upper canine; *c*, lower canine; *P*, upper premolars; *p*, lower premolars; *M*, upper molars; *m*, lower molars.

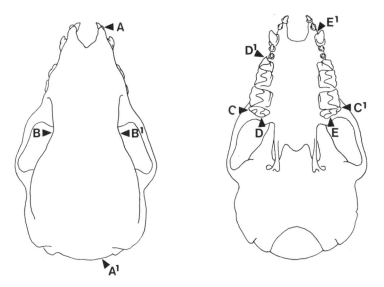

Figure 30. The skull measurements used in key and species descriptions: A-A^1, skull length; B-B^1, postorbital width; C-C^1, width across the last upper molars; D-D^1, length from the last upper premolar to the last upper molar; E-E^1, length of the upper toothrow.

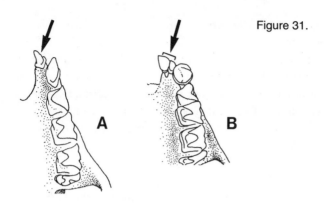

Figure 31.

1a   One upper incisor on each side of skull (Figure 31A) 2
1b   More than one upper incisor on each side of skull (Figure 31B)
.... 4

2a   One upper premolar on each side of skull (Figure 32A)
.... Pallid Bat (*Antrozous pallidus*)
2b   Two upper premolars on each side of skull (Figure 32B) 3

3a   Skull length greater than 15 mm .. Hoary Bat (*Lasiurus cinereus*)
3b   Skull length less than 15 mm
.... Western Red Bat (*Lasiurus blossevillii*)

4a   One upper premolar on each side of skull (Figure 32A)
.... Big Brown Bat (*Eptesicus fuscus*)
4b   More than one upper premolar on each side of skull (Figure 32B)
.... 5

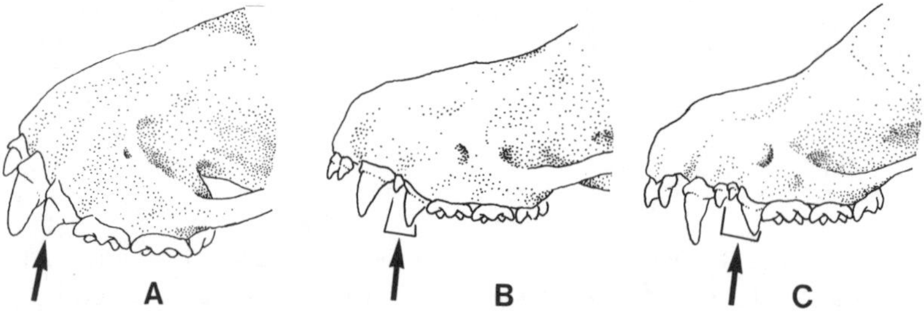

Figure 32.

5a Two upper premolars on each side of skull (Figure 32B) 6
5b Three upper premolars on each side of skull (Figure 32C) 8

6a Skull length greater than 18 mm
.... Spotted Bat (*Euderma maculatum*)
6b Skull length less than 18 mm 7

7a Postorbital width greater than 4 mm
.... Silver-haired Bat (*Lasionycteris noctivagans*)
7b Postorbital width less than 4 mm
.... Townsend's Big-eared Bat (*Plecotus townsendii*)

8a Postorbital width greater than 3.4 mm 9
8b Postorbital width less than 3.4 mm 15

9a Ratio of postorbital width/upper-toothrow-length greater than
0.7 mm 10
9b Ratio of postorbital width/upper-toothrow-length less than 0.7 mm
.... 12

10a Brain-case strongly elevated (Figure 33A)
.... Long-legged Myotis (*Myotis volans*)
10b Brain-case not strongly elevated (Figure 33B) 11

11a Forehead with a steep slope (Figure 34A), skull length usually less
than 14 mm .... Yuma Myotis (*Myotis yumanensis*)
11b Forehead with a gradual slope (Figure 34B), skull length usually
greater than 14 mm ..... Little Brown Myotis (*Myotis lucifugus*)

12a Length across upper molars and last premolar on skull greater than
4.2 mm 13
12b Length across upper molars and last premolar on skull less than
4.2 mm. 14

13a Width across last upper molars greater than 6.2 mm
.... Fringed Myotis (*Myotis thysanodes*)
13b Width across last upper molars less than 6.2 mm
.... Western Long-eared Myotis (*Myotis evotis*)

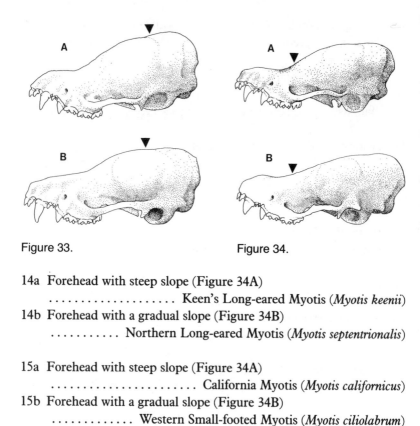

Figure 33.   Figure 34.

14a Forehead with steep slope (Figure 34A)
.... Keen's Long-eared Myotis (*Myotis keenii*)
14b Forehead with a gradual slope (Figure 34B)
.... Northern Long-eared Myotis (*Myotis septentrionalis*)

15a Forehead with steep slope (Figure 34A)
.... California Myotis (*Myotis californicus*)
15b Forehead with a gradual slope (Figure 34B)
.... Western Small-footed Myotis (*Myotis ciliolabrum*)

SPECIES ACCOUNTS

A detailed species account is provided for each of the province's 16 vespertilionid bat species. The Big Free-tailed Bat (*Nyctinomops macrotis*), a molossid bat, is not included in the accounts. The only record of this bat in British Columbia is of a specimen that was collected in November 1938 at Essondale, New Westminster. This record undoubtedly represents an accidental occurrence, because the nearest known breeding populations are found in Utah and this bat often turns up outside its normal range during the autumn migration period.

Information in the species accounts is divided into six categories:

Other Common Names is a list of alternate English common names.

Description is a concise description including measurements, dental formula and a comparison with similar species that may be confused in identification. Most of our descriptions of fur colour, body measurements and weights are based on adult museum specimens and live animals from British Columbia. Exceptions are the Northern Long-eared Myotis (*Myotis septentrionalis*), Western Red Bat (*Lasiurus blossevillii*), Spotted Bat (*Euderma maculatum*) and Pallid Bat (*Antrozous pallidus*). Because we have only a few museum specimens and records of live captures in British Columbia, we had to base their descriptions on individuals from other parts of North America.

All linear measurements are in millimetres and weight is in grams. The values given are the mean, range (in parenthesis) and number of individuals measured. For example, the mean forearm length of the California Myotis (*Myotis californicus*) is 32.1 mm, based on measurements obtained from 68 animals with forearm lengths ranging from 29.4 to 35.0 mm—these figures are written as "32.1 (29.4-35.0) n = 68".

Because the weight of bats varies with season and reproductive condition, there is considerable variation in the mean for each species.

The dental formula describes the number of each kind of tooth in one side of the head (Figure 29). The first figure is the number in the upper jaw; the second figure is the number in the lower jaw. For example, incisors 1/1, canines 1/1, premolars 2/3, molars 3/3 indicates one upper and lower incisor, one upper and lower canine, two upper and three lower premolars, and three upper and lower molars.

Natural History includes habitat, roosting sites, hibernation sites, food habits, behaviour and reproduction. Wherever possible, we used information from studies done on British Columbia populations. Considerable habitat, elevational and reproductive data were also obtained from museum specimens. Nevertheless, detailed studies on the province's bats are limited to a few species in the Okanagan Valley. Therefore, for many species we had to rely on information from research done in other parts of western Canada or the United States. Furthermore, it should be noted that there is great variation in the amount of information available for our bat species. The Little Brown Myotis (*Myotis lucifugus*) is probably the most studied of any North American bat and its natural history account is relatively comprehensive. In contrast, almost nothing is known about the general biology of Keen's Long-eared Myotis (*Myotis keenii*) and its account is sketchy.

Range is a general description of the overall distribution and a detailed description of the range in the province. It includes a range map that is based on an extensive review of all the known locality records from the province. The data base for the maps consisted of 1700 museum specimens and 500 observational records. Detailed information on specific records can be obtained from the Royal British Columbia Museum. Maps with symbols representing actual locality records were generated by computer using Universal Transverse Mercator Grid (UTM) co-ordinates. Each dot represents a record of occurrence substantiated by a specimen or observational data. Provincial base maps (transverse mercator projections) are all 1:2,000,000 scale. For species with very localized ranges in the province, inset maps depicting ranges on a small scale are added to the provincial base map. Some of the biases in the species range maps are shown in Figure 35, which is derived from all known bat records from British Columbia. Records are concentrated in southeastern Vancouver Island, the lower Fraser Valley

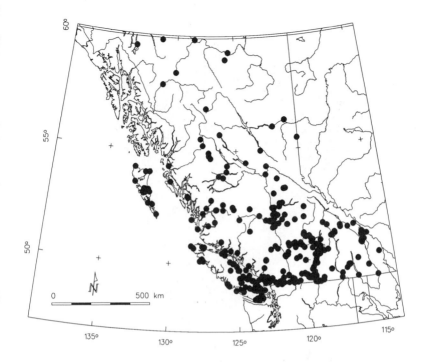

Figure 35. There are 2,200 locality records for the 16 species of bats in British Columbia, most from the southern areas of the province.

(including Vancouver), the Okanagan Valley and Williams Lake region. This may reflect the greater diversity of bats in those regions, but it is primarily due to greater activity by museum collectors, naturalists and bat researchers there. In contrast, few records exist from northern British Columbia, the central and northern Coast Mountains and western Vancouver Island, because little survey work has been done in those regions.

Taxonomy lists subspecies found in the province. All lists are based on *The Mammals of British Columbia: A Taxonomic Catalogue*.

Remarks gives interesting details about the species in British Columbia and identifies areas for future study.

Selected References directs the reader to citations of important publications on the species. It is not intended to be a comprehensive review of all literature; the emphasis is on studies done in British Columbia and adjacent areas in western North America.

California Myotis
Myotis californicus

California Myotis
Myotis californicus

Other Common Names: California Bat.

Description

The California Myotis is one of our smallest species. Its fur colour varies from rusty to blackish brown and is dull, lacking a glossy sheen. Its ears and its wing and tail membranes are black. The ears extend beyond the nose when pushed forward; the tragus is long and narrow. The length of the naked area on the snout is roughly equal to the width across the nostrils. The hind foot is small; the calcar has a prominent keel. The small, delicate skull has a steeply sloping forehead.

Measurements:

total length: 80 (65-95) n = 75
tail vertebrae: 36 (26-41) n = 77
hind foot: 6 (5-9) n = 80
ear: 13 (8-15) n = 40
tragus: 4 (4-8) n = 35
forearm: 32.1 (29.4-35.0) n = 68
wingspan: 222 (209-251) n = 28
weight: 4.4 (3.3-5.4) n = 26.

Dental Formula:

incisors: 2/3
canines: 1/1
premolars: 3/3
molars: 3/3

Identification:

In south-central British Columbia, the California Myotis can be confused with the West-

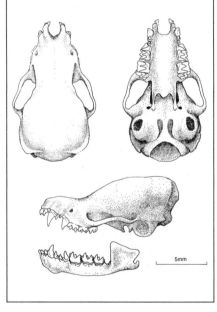

ern Small-footed Myotis (*Myotis ciliolabrum*), but the latter has paler fur that contrasts more sharply with the black wings, face and ears, and it has a longer bare area on its snout. In other parts of the province the small size (forearm less than 36 mm), small foot and keeled calcar will readily distinguish the California Myotis from other *Myotis* species. The small skull (less than 3.4 mm postorbital width) separates this bat from all other species of *Myotis* except the Western Small-footed Myotis. The skull of the California Myotis has a more steeply sloped forehead than the skull of the Western Small-footed Myotis.

Natural History

In British Columbia, this species inhabits arid grasslands, humid coastal forests and montane forests. Its elevational range is from sea level on the coast to 1280 metres in Glacier National Park. It uses rock crevices, tree cavities, spaces under the bark of trees, mine tunnels, buildings and bridges for summer day roosts. Small maternity colonies have been found in similar locations. This bat is particularly flexible in its choice of night roosts and it will use almost any natural or man-made shelter. Males live separately from females in summer and often appear to change their roosting locations. Studies in the Okanagan Valley revealed that the California Myotis emerges shortly after dusk. There are two peaks in hunting activity: between 10:00 and 11:00 pm and between 1:00 and 2:00 am. It hunts mostly over lakes near the surface, although this species will forage in the tree canopy, especially in poplar groves. In the Okanagan, the diet is predominately caddisflies, with moths, flies and beetles minor food items. Food habits in other parts of the province have not been studied.

There are several winter records for the coast, from Vancouver Island and in the Vancouver area. One Vancouver record was of an individual flying inside a building in January at the University of British Columbia. Insect remains were present in its stomach suggesting that it had recently fed. In coastal Washington and Oregon, where the California Myotis frequently hibernates in buildings, there is considerable evidence that it occasionally emerges from torpor to feed. No hibernating colonies have been recorded in the interior of the province, but there are winter museum specimens from Hope (January), Okanagan Landing (March) and Rogers Pass, Glacier National Park (January).

Mating occurs in autumn; females produce a single young. The breeding biology of this bat in British Columbia is poorly documented. There are records of pregnant females in the interior from 11 June to 26 June and on Saltspring Island on 6 June, suggesting that young are born in late June or early July.

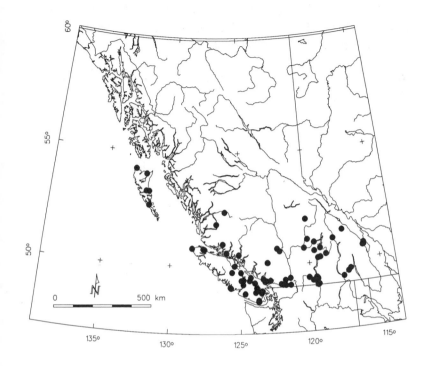

Range

A western bat, its range extends from southern Mexico to British Columbia and southeastern Alaska. In British Columbia, it inhabits several coastal islands including Vancouver Island and the Queen Charlotte Islands, the coastal mainland north to the Bella Coola Valley, and the interior north to Wells Gray Provincial Park and east to Kootenay National Park.

Taxonomy

Two subspecies are found in the province: *M. c. caurinus*, a dark coastal race ranging from California to Alaska, and *M. c. californicus*, a paler race that inhabits the western United States and the southern interior of British Columbia.

Remarks

This bat is common in British Columbia. Although it has been studied in the Okanagan Valley, its biology in other regions, particularly the humid coastal forest, is virtually unknown.

Selected References: Cowan (1942), Krutzsch (1954), Woodsworth (1981).

Western Small-footed Myotis
Myotis ciliolabrum

Other Common Names: Western Small-footed Bat.

Description

The Western Small-footed Myotis is the smallest bat in British Columbia. Its fur colour varies from pale tan to orange-yellow on the back; the underside is paler, almost buff. The black ears, wing and tail membranes, face and snout contrast strikingly with the pale fur. The length of the naked area on the snout is about one and a half times the width across the nostrils. The ears are relatively long; the tragus is long and narrow. There is a distinct keel on the calcar; the foot is small. The skull is very small and delicate with a gradually sloping forehead.

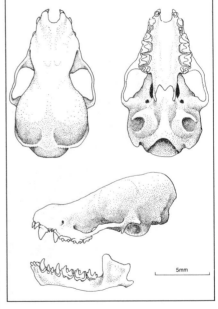

Measurements:
  total length: 83 (72-90) n = 44
  tail vertebrae: 38 (32-45) n = 44
  hind foot: 7 (6-8) n = 44
  ear: 13 (8-15) n = 25
  tragus: 7 (4-9) n = 24
  forearm: 31.8 (28.8-33.4) n = 47
  wingspan: 221 (205-245) n = 25
  weight: 4.6 (2.8-5.5) n = 23

Dental Formula:
  incisors: 2/3
  canines: 1/1
  premolars: 3/3
  molars: 3/3

Identification:
This species is smaller than the Fringed Myotis (*Myotis thysanodes*) and the Long-legged Myotis (*Myotis volans*). Its small foot and presence of a prominent keel on the calcar distinguishes it from the Western Long-eared Myotis (*Myotis evotis*), Little Brown Myotis (*Myotis lucifugus*) and Yuma Myotis (*Myotis yumanensis*). The California Myotis (*Myotis californicus*)

Western Small-footed Myotis
Myotis ciliolabrum

is similar in size and possesses a keeled calcar; see that species account for distinguishing traits.

Natural History

The Western Small-footed Myotis lives near cliffs and rock outcrops in arid valleys and badlands. Its elevational range in British Columbia is from 300 to 850 metres. In summer it roosts in cavities in cliffs, boulders, vertical banks, the ground and talus slopes, and under rocks. It prefers small, protected crevices where the environment is dry and hot (27-33°C). Nursery colonies are situated in similar sites (although in California a mixed colony of males and pregnant females was discovered under the wallpaper of an abandoned house). Small caves, abandoned mine adits and buildings serve as night roosts.

In the Okanagan the Western Small-footed Myotis feeds primarily on caddisflies and also eats other flies, beetles and moths. It hunts over the edge of rocky bluffs and rarely over open water. There are two peaks in feeding activity: between 10:00 and 11:00 pm and between 1:00 and 2:00 am.

The Western Small-footed Myotis hibernates in winter. In Idaho and Montana, it has been found hibernating in caves and abandoned mines where it prefers temperatures of 1.5°-5.5°C. Recent observations indicate that a few individuals overwinter in British Columbia. In the Okanagan and Williams Lake regions, small numbers (1-3) have been found hibernating in caves and mine adits. Western Small-footed Myotis usually hibernate alone, wedging themselves into tight crevices or depressions in the ceiling of the hibernaculum with their undersides pressed against the ceiling and their heads facing outwards. This seems to provide the most stable environment for them.

Mating probably takes place in autumn before hibernation. Females usually have a single young, although there is a record of twins from South Dakota. In British Columbia, pregnant females have been observed from 21 June to 13 July and nursing females from 13 June to 3 August. This suggests that young are born from mid-June to mid-July. Newborn young weigh about one gram and have a forearm length of 12 millimetres.

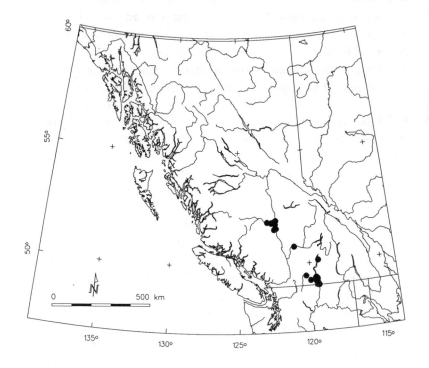

Range

This species occurs throughout most of western North America, where it ranges from Mexico to southern Canada in Saskatchewan, Alberta and British Columbia. In British Columbia, it occurs in the dry interior valleys as far north as the Chilcotin River and Williams Lake region.

Taxonomy

In Cowan and Guiguet's *The Mammals of British Columbia* this species was classified as *Myotis subulatus*; however, in most of the literature the Small-footed Myotis is classified as *Myotis leibii*. Western forms of the Small-footed Myotis are now treated as a distinct species, *Myotis ciliolabrum*. All British Columbia populations belong to the subspecies *M. c. melanorhinus*, a golden-brown race that ranges from Mexico to British Columbia.

Remarks

The Western Small-footed Myotis is on the provincial Blue List. Until recently, most biologists thought that in British Columbia this species was restricted to the Okanagan and Similkameen valleys. However, recent bat survey work in the dry interior has revealed that this bat is more widespread.

Selected References: Genter (1986), Tuttle and Heaney (1974), Woodsworth (1981).

Western Long-eared Myotis
Myotis evotis

Western Long-eared Myotis
Myotis evotis

Other Common Names: Long-eared Myotis, Long-eared Bat.

Description

The Western Long-eared Myotis is a large *Myotis* species with long ears extending 5 mm or more beyond the nose when pushed forward. Its dorsal fur colour is extremely variable, ranging from yellowish brown in the interior of the province to dark brown or nearly black in coastal areas. Blackish brown shoulder patches are usually evident but they can be indistinct on dark individuals. The outer edge of the tail membrane has a fringe of tiny hairs that can be seen with a hand lens. The ears and flight membranes are nearly black and usually contrast sharply with the paler fur. The tragus is long and slender with a small lobe at its base. The calcar lacks a distinct keel. The skull has a relatively broad interorbital area and a gradually sloping forehead.

Measurements:
total length: 92 (74-103) n = 54
tail vertebrae: 42 (31-50) n = 51
hind foot: 9 (7-11) n = 51
ear: 20 (17-22) n = 31
tragus: 10 (8-12) n = 21
forearm: 38.4 (36.0-42.0) n = 47
wingspan: 271 (243-294) n = 21
weight: 5.5 (4.2-8.6) n = 25

Dental Formula:
incisors: 2/3
canines: 1/1
premolars: 3/3
molars: 3/3

Identification:
Three other long-eared *Myotis* species are found in British Columbia: Keen's Long-eared

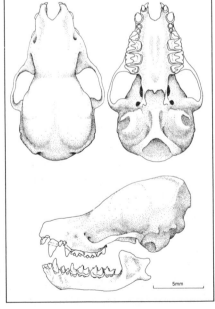

Myotis (*Myotis keenii*), the Northern Long-eared Myotis (*Myotis septentrionalis*) and the Fringed Myotis (*Myotis thysanodes*). All have ranges that overlap with that of the Western Long-eared Myotis. The Fringed Myotis can be readily distinguished by the fringe of hairs on the tail membrane and by the longer forearm (greater than 40 mm). To distinguish the Western Long-eared Myotis from Keen's Long-eared Myotis and the Northern Long-eared Myotis using external features see their species accounts. The skull of the Western Long-eared Myotis has a longer toothrow than that of Keen's Long-eared Myotis and the Northern Long-eared Myotis: the distance from the last upper premolar to the last upper molar is greater than 4.2 mm.

Natural History

This species is found in a wide range of habitats in the province, from arid grasslands and ponderosa pine forests to humid coastal and montane forests. Its elevational range in British Columbia extends from sea level on the coast to 1220 metres in the Cascades (Manning Provincial Park) and Rocky Mountains (Kootenay National Park). This is one of the few bats found consistently at high elevations in western Canada. In the Kananaskis region of the Alberta Rockies, summer populations consisting of both sexes live at 1350 to 2050 metres. In summer the Western Long-eared Myotis uses buildings or under the bark of trees as day roosts; there are also a few records of this species roosting in caves, sink holes and fissures in cliffs. Maternity colonies, usually located in buildings, are generally small (5 to 30 individuals) and may contain a few adult males. Caves and mine adits are used as temporary night roosts.

Western Long-eared Myotis typically emerge 10 to 40 minutes after dark to feed. It is quite flexible in its feeding behaviour, eating airborne insects as well as gleaning insects from vegetation or off the ground. Food habits have not been studied in British Columbia; in other regions, it is known to prey mainly on moths, as well as beetles, flies and spiders. Robert Barclay has suggested that bat's flexible feeding behaviour enables females to breed successfully in high, cool sites where flying insects are scarce. Its quiet, short-duration, high-pitched echolocation calls are an adaptation for hunting in habitats with heavy vegetation. Furthermore, these calls are not readily detected by most moths. When closing in for an attack, the Western Long-eared Myotis often stops calling and listens for sounds produced by its prey. Moths may be especially vulnerable to predation by the Western Long-eared Myotis because their fluttering is audible.

There are no winter records for the province; in fact, this bat's winter biology is poorly documented throughout its range. In the western United States, a few individuals have been found hibernating in caves and mine adits and there is a December record from coastal Oregon of an individual that was found in a garage. Reproductive data for the province are also scanty, consisting of anecdotal observations and information on museum specimens. Mating presumably occurs in autumn or early winter; females produce a single young. Pregnant females were found in a maternity colony at Vermilion Crossing, Kootenay National Park, between 28 June and 7 July. In the Okanagan Valley, females with near-term foetuses were collected on 27 and 28 June, and nursing females were observed from 5 July to 13 August. These data suggest

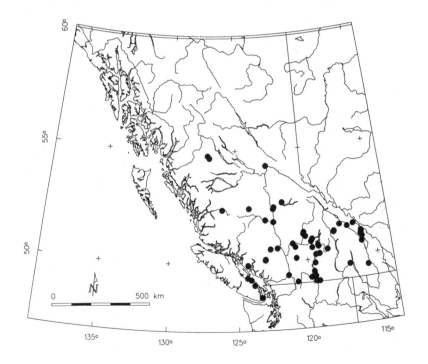

that, in the interior, young are born in late June or early July. No reproductive data are available for coastal British Columbia. Newborn Western Long-eared Myotis are furless and weigh 1.0-1.5 grams.

Range

This is a western bat that ranges from Baja California through the western United States to Saskatchewan, Alberta and British Columbia, where it occupies the entire mainland as far north as the Bella Coola Valley on the coast and the Prince George region in the interior. It also inhabits Vancouver Island but appears to be absent from the Queen Charlotte Islands.

Taxonomy

Two subspecies occur in the province: *M. e. pacificus*, a dark coastal race ranging from California as far north as the Bella Coola Valley in British Columbia, and *M. e. evotis*, a paler, larger race that inhabits Mexico, the western United States and western Canada. In British Columbia, it is found in the south-central interior north to the Prince George region.

Remarks

Despite its widespread distribution in the province, remarkably little is known about the basic biology of this species particularly in the coastal forests. A published record from the Okanagan reports that a dead Western Long-eared Myotis was found in the mouth of a snake, a Western Yellow-bellied Racer.

Selected References: Barclay (1991), Faure et al. (1990), Manning and Jones (1989), Maser et al. (1984).

Keen's Long-eared Myotis
Myotis keenii

Other Common Names: Keen Bat, Keen's Long-eared Bat, Keen's Myotis.

Description

Keen's Long-eared Myotis is a medium-sized *Myotis* species with dark, glossy fur and darker indistinct spots on the back of the shoulder; the underside is paler. Its long ears extend beyond the nose when pressed forward; the tragus is long, narrow and pointed. The ears and wing membranes are dark brown but not black. The outside edge of the tail membrane has a fringe of tiny hairs that can be seen with a hand lens. The calcar has an indistinct keel. The skull has a relatively steep forehead.

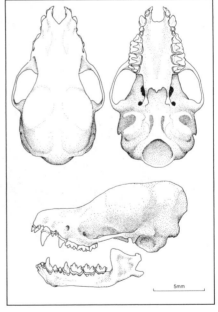

Measurements:
  total length: 84 (63-94) n = 33
  tail vertebrae: 40 (32-44) n = 33
  hind foot: 9 (8-10) n = 32
  ear: 18 (16-20) n = 38
  tragus: 11 (9-12) n = 41
  forearm: 36.0 (34.2-38.5) n = 43
  wingspan: 241 (209-262) n = 33
  weight: 5.1 (4.0-5.9) n = 35

Dental Formula:
  incisors: 2/3
  canines: 1/1
  premolars: 3/3
  molars: 3/3

Identification:
The Fringed Myotis (*Myotis thysanodes*), Northern Long-eared Myotis (*Myotis septen-trionalis*) and Western Long-eared Myotis (*Myotis evotis*) are other *Myotis* species with a long ear and tragus. However, the distribution of Keen's Long-eared Myotis overlaps only with that of the Western

Keen's Long-eared Myotis
Myotis keenii

Long-eared Myotis. No external features that we know of will positively distinguish the two species. Keen's Long-eared Myotis has relatively shorter ears, extending less than 5 mm beyond the nose when pushed forward, and its ears are also paler; but these traits are variable and somewhat unreliable. Positive identification can only be made from cranial characters; the distance from the last upper pemolar to the last upper molar is less than 4.2 mm in Keen's Long-eared Myotis and greater than 4.2 mm in the Western Long-eared Myotis.

Natural History

What little is known about the biology of this bat is derived from incidental observations, information recorded on museum specimens and some research conducted on Hot Spring Island in the Queen Charlotte Islands. The distributional pattern suggests that Keen's Long-eared Myotis is associated with coast forest habitats. It likely uses tree cavities, rock crevices and small caves as roosting sites. There are several records from the vicinity of hot springs and the only known colony occurs on Hot Spring Island. Most of the natural history data for this species comes from the Hot Spring Island population. However, given the unusual ecological situation, it is not clear if this information is applicable to other populations. On Hot Spring a colony of about 70 Keen's Long-eared Myotis, in association with a colony of the Little Brown Myotis, roosts under rocks that are heated by a natural hot spring (Figure 11). Temperatures at the roost entrance in summer range from 22°C to 27°C, whereas the ambient temperature varies from 11°C to 18°C. Because of a warm spring, the roost is quite humid. The roost is situated below the high tide line and it is often submerged for several hours at high tide. During these periods of high tides the roost is abandoned.

Food habits have not been investigated, but the diet probably consists of moths and other insects. On Hot Spring Island, Keen's Long-eared Myotis forages over hot spring pools and clearings above salal. Individuals marked with light tags were observed to fly into the tree canopy, but their activity could not be monitored. Mating presumably occurs in autumn. The scanty breeding data consist of a nursing female and juveniles observed on Hot Spring Island in late July; this suggests that the young are born in June or early July. There are no winter records, and it is unknown if this species hibernates in coastal regions.

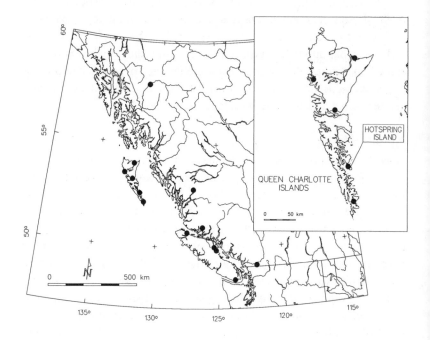

Range

Keen's Long-eared Myotis is the only North American bat restricted to the Pacific coastal region. The few locality records available suggest that its range extends from the Olympic Peninsula in Washington to southeastern Alaska. In British Columbia, it is found on the coastal mainland as far north as the Stikine River, on the Queen Charlotte Islands and on Vancouver Island.

Taxonomy

No subspecies are recognized. In much of the earlier literature, two subspecies were recognized for this bat: *M. k. keenii* from the coast and *M. k. septentrionalis* from central and eastern North America. The latter is now considered to be a distinct species *Myotis septentrionalis*.

Remarks

Most of the population of Keen's Long-eared Myotis appears to be restricted to British Columbia. In fact, of the known locality records only three are from outside the province: one from Wrangell Island in southeastern Alaska and two from the Olympic Penninsula in Washington. Specimens of Keen's Long-eared Myotis reported from Stuie, Telkwa and Parksville in previous publications were misidentified; they are the Western Long-eared Myotis.

In response to its apparent rarity and the lack of knowledge about its basic biology, the provincial Ministry of Environment placed Keen's Long-eared Myotis on the provincial Red List. This bat is rare in museum collections, being represented by a small number of specimens from only nine locations in the province, but there is not enough information to determine if it is rare in nature. This species may be common in coastal forests but seldom captured because of its foraging or roosting behaviour.

At present, too little is known about this bat to speculate on its use of old-growth forest habitat. Additional field studies are urgently required to determine the status of this species. Until a reliable field technique is developed to distinguish live Keen's Long-eared Myotis from the Western Long-eared Myotis, field research in areas where the two species co-exist is virtually impossible. Because Keen's Long-eared Myotis may be an endangered species, killing animals to identify from cranial traits is clearly unacceptable. Studies are underway to develop a set of external measurements that will separate the two species in the hand.

Selected References: van Zyll de Jong (1979).

Little Brown Myotis
Myotis lucifugus

Little Brown Myotis
Myotis lucifugus

Other Common Names: Little Brown Bat.

Description

The Little Brown Myotis is a medium-size species of *Myotis*. Its fur colour is extremely variable: the fur on its back ranges from yellow or olive in populations from the dry interior to blackish in coastal populations (see inside front cover). The fur on its underside is lighter, varying from light brown to tan. Its dorsal fur is long and glossy. The wing membranes and ears are dark brown. The ears reach the nostrils when pushed forward; the tragus is blunt and about half the ear length. The calcar is not keeled. The skull is typical of most myotis species; the forehead has a gradual slope.

Measurements:

total length: 86 (70-108) n = 383
tail vertebrae: 37 (25-59) n = 379
hind foot: 10 (6-13) n = 385
ear: 13 (9-17) n = 217
tragus: 7 (4-10) n = 151
forearm: 36.4 (33.0-40.3) n = 295
wingspan: 248 (224-274) n = 151
weight: 6.2 (6.2-10.2) n = 98

Dental Formula:

incisors: 2/3
canines: 1/1
premolars: 3/3
molars: 3/3

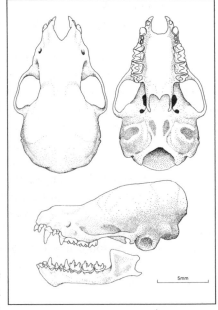

Identification:

The ear and tragus of the Little Brown Myotis are smaller than that of the Fringed Myotis (*Myotis thysanodes*), Northern Long-eared Myotis (*Myotis septentrionalis*), Western Long-eared Myotis (*Myotis evotis*) and Keen's Long-eared Myotis (*Myotis keenii*). The absence of a keel on the calcar distinguishes it from the Long-legged Myotis (*Myotis volans*), California Myotis (*Myotis californicus*) and Western Small-footed Myotis (*Myotis ciliolabrum*). The Yuma Myotis (*Myotis yumanensis*) is smaller (forearm usually less than 36 mm) with duller, shorter fur. Nevertheless, in some parts of the province, such as the Okanagan Valley, the Little Brown Myotis can be extremely difficult to distinguish from the Yuma Myotis. Behaviour may assist in identification. The Little Brown Myotis tends to be more aggressive than the Yuma Myotis when handled. Skull features will give a more reliable identification: Little Brown Myotis has a longer skull (greater than 14 mm) that has a more gradually sloping forehead; but even these characteristics are not always reliable in identifying some individuals with intermediate skull features.

Natural History

In British Columbia this bat exploits a wide range of habitats, from arid grassland and Ponderosa Pine forest to humid coastal forest and northern boreal forest. With records from sea level on the coast to 2288 metres above sea level in the Rocky Mountains (Mount Assiniboine Provincial Park), the Little Brown Myotis has the greatest altitudinal range of any of our bats. Robert Barclay's research in the Alberta Rockies indicates that males are more abundant than females at higher elevations. Similarly, at 300 to 600 metres elevation in the Cascades of western Washington, Donald Thomas observed few females and none were reproductively active.

Summer roosts are in buildings and other man-made structures, tree cavities, rock crevices, caves and under the bark of trees. In summer the sexes live separately. Females congregate in nursery colonies that may contain hundreds or even thousands of individuals. Nursery colonies are in sites with hot (30-55°C), stable temperatures. Nursing females seem to prefer attics, but they will roost in almost any natural site that offers the environmental conditions that will promote the rapid growth of young. For example, a nursery colony of several hundred Little Brown Myotis was discovered in a small cave on the Grayling River in northern British Columbia. A hot (30°C), humid environment is maintained inside this cave by a natural hot spring.

Males rarely occupy nursery colonies. In summer they can be found roosting alone or in small colonies usually in sites that are cooler than the nurseries.

The Little Brown Myotis emerges at dusk to feed. It is an opportunistic hunter that eats a great variety of insect prey. Robert Herd's research in the southern Okanagan Valley is the only study of food habits done in the province. He found that the Little Brown Myotis hunts in the valley and nearby hills in Ponderosa Pine forests, openings of trees, and over bluffs, lakes, rivers and irrigation flumes. Aquatic insects such as midges, caddisflies and mayflies are the major prey, although beetles, moths and other kinds of flies are also taken. The diet changes seasonally in response to insect abundance, with midges predominant in spring and caddisflies and mayflies most important in summer. The Little Brown Myotis is able to adjust its hunting techniques quickly to take advantage of insect concentrations. Most prey is captured in the air and eaten while flying. After an initial feeding period of 15 to 20 minutes, individuals occupy temporary night roosts near the day retreat. Night roosts are used most often when temperatures are cool (below 15°C)—a protected night roost may help the Little Brown Myotis remain active so that it can digest its meal faster.

The Little Brown Myotis hibernates in caves and abandoned mines; it does not appear to hibernate in buildings. Some hibernacula in eastern Canada contain thousands of individuals. Hibernation records in British Columbia are limited to several old mines in the interior, each containing a few individuals. The whereabouts of most of the British Columbian populations in winter is unknown. Banding studies indicate that this bat will migrate 50 to 200 kilometres between hibernacula and summer roosts and if undisturbed it occupies the same sites year after year.

Before entering hibernation there is a period in late summer and early autumn when adults and young of the year make nocturnal flights through caves and mines, sites that are potential hibernacula. This swarming behaviour begins with the breakdown of the nursery colonies in late July and lasts until the bats enter hibernation. The role of swarming is not clear. It may be to familiarize young bats with the locations of potential hibernacula. Hibernation begins in September or October depending on the local climate. The Little Brown Myotis selects areas in the hibernaculum where there is high humidity (70-95%) and the temperature is above freezing (1-5°C). It may hibernate alone or in clusters. A change in temperature will waken this bat from hibernation to search for a new location in the hibernaculum. There are a few records of individuals moving to a new hibernaculum in mid winter. While hibernating, the Little Brown Myotis steadily burns up its fat reserves and loses about 25% of its weight by spring. In the interior of the province, hibernation probably lasts until April or early May, but in coastal regions this bat may arouse in late winter—it has been found feeding in coastal Oregon in mid March.

Mating occurs in late summer and early autumn before entering hibernation. Males do not breed in their first autumn. Although yearling females can breed in their first year, in British Columbia most delay reproduction until their second autumn. After a gestation period of 50 to 60 days, a single young is born; twins are extremely rare. Birth dates vary, determined in part by the date when females arouse from hibernation. In the southern Okanagan Valley, females give birth between the second week of June and the second week of July. Populations living at higher elevations and latitudes probably have their young later. There are no data on birth dates for coastal areas, but the earliest pregnancy record is 4 June from eastern Vancouver Island. Newborn young weigh 1.0-1.5 grams. They grow rapidly and by three weeks are capable of flying and eating solid food. Banding studies have revealed that this bat has a long life span with some individuals living more than 30 years.

Range

A widespread species, the Little Brown Myotis inhabits most of North America as far north as the tree-line. In British Columbia it is found throughout the entire mainland and on several islands, including Vancouver Island and the Queen Charlotte Islands.

Taxonomy

Three races occur in the province. *M. l. alascensis*, a dark subspecies that ranges from California to southeastern Alaska, is found throughout most of the province. *M. l. carissima* is a pale race that inhabits part of the western United States, southern Alberta and the southern dry interior of British Columbia. *M. l. lucifugus*, a widespread race found across eastern and central North America, is restricted to extreme northern British Columbia.

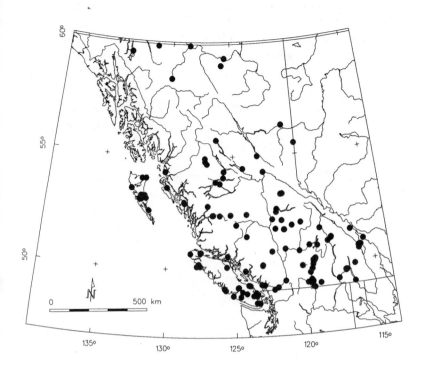

Remarks

Although the Little Brown Myotis is the most widespread and one of the most abundant bats in the province, there is still much to be learned about the basic biology of this species.

Selected References: Aldridge (1986), Fenton (1970), Fenton and Barclay (1980), Herd and Fenton (1983), Humphrey and Cope (1976), Schowalter et al. (1979).

Northern Long-eared Myotis
Myotis septentrionalis

Northern Long-eared Myotis
Myotis septentrionalis

Other Common Names: Northern Long-eared Bat, Northern Myotis.

Description

The Northern Long-eared Myotis is a medium-size species of *Myotis* with dark brown fur on its back. The fur on its underside is paler, varying from tawny to pale brown. The ears and flight membranes are dark brown but not black. The ears are long, extending beyond the nose when pushed forward; the tragus is long, narrow and pointed. The edge of the tail membrane is bare or has only a few, scattered hairs. The calcar may have an indistinct keel. The skull is relatively narrow with a long rostrum.

Measurements:
total length: 87 (80-94) n = 18
tail vertebrae: 39 (29-46) n = 63
hind foot: 9 (7-11) n = 61
ear: 17 (14-19) n = 56
tragus: 10 (8-12) n = 45
forearm: 36.1 (34.0-38.0) n = 57
wingspan: 234 n = 1
weight: 6.5 (5.0-10.0) n = 33

Dental Formula:
incisors: 2/3
canines: 1/1
premolars: 3/3
molars: 3/3

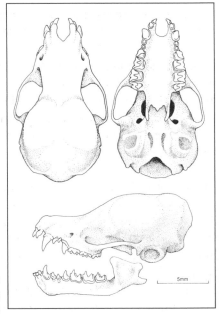

Identification:
The range of the Northern Long-eared Myotis overlaps with only one other long-eared *Myotis* species in British Columbia—the Western Long-eared Myotis (*Myotis evotis*). The Northern Long-eared Myotis can be discriminated by its paler and smaller ears (extending less than 5 mm beyond the

nose), the lack of shoulder spots and the sparse fringe of hairs on the outer edge of the tail membrane. Its skull can be distinguished from that of the Western Long-eared Myotis by its shorter toothrow—the distance from the last upper premolar to the last upper molar is less than 4.2 mm.

Natural History

In Canada the Northern Long-eared Myotis is generally associated with boreal forests. Information on habitat in British Columbia is limited to Mount Revelstoke National Park where it has been found in Western Hemlock - Western Red-cedar forests at about 700 metres elevation. No roosts have been found in the province. In eastern North America summer day roosts and nursery colonies have been found in buildings and under the bark of trees; all nursery colonies were small, comprising no more than 30 individuals. Caves may be exploited for temporary night roosts.

This species emerges at dusk to hunt over small ponds and forest clearings under the tree canopy. In Mount Revelstoke National Park Northern Long-eared Myotis were observed regularly at dusk drinking from small pools in forest clearings. Much of this species' hunting activity takes place just above the understory one to three metres above the ground. Its diverse diet includes caddisflies, moths, beetles, flies and leafhoppers. Some prey may be gleaned from twigs and foliage.

There are no winter records for the province. However, in other parts of North America, the Northern Long-eared Myotis hibernates in caves and abandoned mine tunnels. Swarming behaviour begins in late summer or early autumn and movements of up to 56 kilometres between the summer roost and the hibernaculum have been documented. This species appears to be a late hibernator; in eastern Canada it arrives at hibernacula two to eight weeks after the Little Brown Myotis enters hibernation. The Northern Long-eared Myotis hibernates alone or in small clusters, selecting tight crevices or drill holes where temperatures may be as cool as 1.6°C. Although this species and the Little Brown Myotis often share hibernacula, they are rarely found touching while hibernating.

Mating usually takes place at the hibernaculum in autumn; females produce a single young. The only available breeding datum for British

Columbia, a female with a six-millimetre embryo found on 16 June, suggests that young are born in late June or early July.

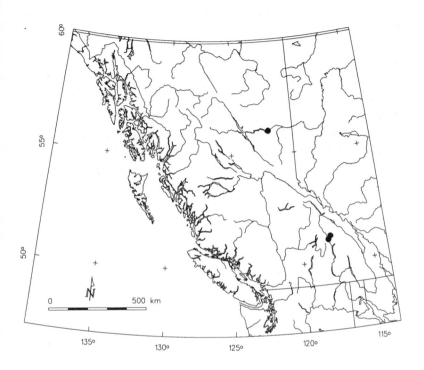

Range

This bat is distributed across the eastern United States and Canada west to the southern Northwest Territories and eastern British Columbia. There are only three substantiated locality records for the province: Hudsons Hope in the Peace River area, Mount Revelstoke National Park and the Revelstoke Dam. It might also occur in Glacier and Kootenay national parks: echolocation calls resembling those of Northern Long-eared Myotis were recorded in both parks during bat surveys with electronic bat detectors; but echolocation calls of the Northern Long-eared Myotis and the Western Long-eared Myotis cannot be discriminated with certainty using a narrow band detector.

Taxonomy

No subspecies are recognized. In earlier literature this species was classified as a race of *Myotis keenii*.

Remarks

One of the rarest bats in the province, the Northern Long-eared Myotis is on the provincial Red List. The first record for this mammal in British Columbia was collected at Hudsons Hope in 1931. In June 1979, a second specimen was collected on the Giant Cedars Trail in Mount Revelstoke National Park. In subsequent field work in 1981 and 1982 more individuals of this species were trapped and released at this site. An individual was found dead in the powerhouse of the Revelstoke Dam in 1980. With records as far north as Nahanni National Park in the Northwest Territories, it seems likely that the Northern Long-eared Myotis is a regular inhabitant of the extreme northeastern region of the province.

Selected References: Caire et al. (1979), Fenton et al. (1983), van Zyll de Jong et al. (1980).

Fringed Myotis
Myotis thysanodes

Other Common Names: Fringed Bat.

Description
The Fringed Myotis is one of the largest *Myotis* species in British Columbia. Its dorsal fur is pale brown; the fur on its underside is much paler. The ears are long, extending well beyond its nose when pushed forward; the tragus is long and slender. The calcar is not keeled. The outer edge of the tail membrane has a distinct fringe of small stiff hairs that can be easily seen with the naked eye. The skull is relatively large and broad compared with most of our *Myotis* species.

Measurements:
total length: 88 (78-93) n = 10
tail vertebrae: 40 (35-44) n = 10
hind foot: 10 (8-11) n = 10
ear: 18 (16-20) n = 10
tragus: 9 (8-11) n = 4
forearm: 42.2 (40.0-44.5) n = 36
wingspan: 279 (250-295) n = 31
weight: 7.1 (5.4-8.4) n = 35

Dental Formula:
incisors: 2/3
canines: 1/1
premolars: 3/3
molars: 3/3

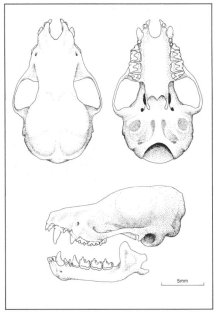

5mm

Identification:
The Long-legged Myotis (*Myotis volans*) and the Western Long-eared Myotis (*Myotis evotis*) are the only two *Myotis* species in the province that are as large as the Fringed Myotis. The Long-legged Myotis has a prominent keel on its calcar, fur on the underwing extending to the knee and elbow and lacks a fringe of hairs on the outer edge of the tail membrane; the Western Long-eared Myotis has longer ears, longer fur and a fringe of

Fringed Myotis
Myotis thysanodes

hairs on the edge of the tail membrane that can only be observed with a hand lens. The skull of the Fringed Myotis is distinguished from that of other *Myotis* species by its narrow postorbital area and the greater distance across the upper molars (more than 6.2 mm).

Natural History

In the western United States the Fringed Myotis typically inhabits desert, arid grassland and arid forest habitat, although it also occurs in the coniferous forests of coastal Oregon. The British Columbian population is associated with arid grassland and Ponderosa Pine - Douglas-fir forest. Its elevational range in the province is 300 to 800 metres.

The Fringed Myotis is a colonial bat that roosts in tightly packed clusters, although considerable movement takes place in the roost in response to temperature changes. With the exception of a nursery colony found in the attic of a house in Vernon in 1937, virtually nothing is known about the roosting habits of the Fringed Myotis in British Columbia. In other parts of its range, caves, mines, rock crevices and buildings are used for day retreats and night roosts. Nursery colonies containing as many as 300 females and their young have been discovered in caves and buildings. In spring and summer, males roost separately and are rarely found in nursery colonies. Small numbers of Fringed Myotis are captured regularly at night in spring and fall at a mine near Oliver. Evidently the mine is used as a temporary night roost because the bat has not been found there during the day. A Fringed Myotis banded at this mine on 6 August 1982 was recaptured at this same site on 7 April 1990.

The Fringed Myotis emerges from its day retreat to hunt one to two hours after sunset. In the Okanagan, this species is often netted in thickets along streams and rivers. It eats moths, flies, beetles, leafhoppers, lacewings, crickets and harvestmen; the presence of flightless insects in the diet indicates that some of its prey is gleaned from foliage.

There are no winter records for the province; in fact there is scant information on the winter biology of the Fringed Myotis throughout its entire range. Accumulation of fat in late summer suggests that this species hibernates, but there are only two reports of hibernating individuals. In the Black Hills of South Dakota, this species has been found hibernating in small numbers in a few caves, and solitary individuals were recently discovered in two caves in Oregon. It appears to hibernate alone clinging to vertical walls with its body upright. There is some circumstantial evidence from the southwestern United States that the Fringed Myotis moves short distances from its summer range to winter hibernation sites.

Detailed data on reproduction for the province are lacking but the nursery colony discovered at Vernon contained juveniles on 19 July suggesting that the young are born in late June or early July. Young of the year males are usually not reproductively active in their first autumn; the age when females reach sexual maturity has not been determined. The young develop quickly and are capable of limited flight when 17 days old; by 3 weeks have they have attained adult size.

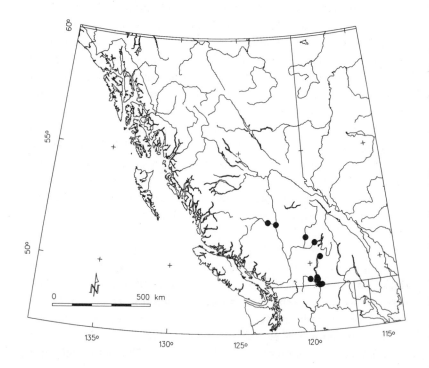

Range

The Fringed Myotis ranges from Mexico to Canada where it is found in the dry interior of British Columbia from Osoyoos to the Chilcotin River and Williams Lake region.

Taxonomy

The population in British Columbia belongs to the subspecies *M. t. thysanodes*, a race that is widely distributed in Mexico and the United States. A coastal subspecies, *M. t. vespertinus*, found in Oregon could reach southwestern British Columbia.

Remarks

The Fringed Myotis is on the provincial Blue List. Although it is now evident that there is a breeding population in south-central British Columbia, this bat's roosting habits, diet and seasonal movements are essentially unknown. Until recently the only records of this species in the province were from the Okanagan Valley. However, recent bat surveys in the dry interior have revealed that this species is more widely distributed.

Selected References: O'Farrell and Studier (1973), O'Farrell and Studier (1980), Martin and Hawks (1972), Maslin (1938).

Long-legged Myotis
Myotis volans

Long-legged Myotis
Myotis volans

Other Common Names: Hairy Winged Bat, Long-legged Bat.

Description
The Long-legged Myotis is one of the largest *Myotis* species in British Columbia. Its fur colour varies from reddish brown to nearly black. The hair on its belly extends to the undersides of the wing membranes as far as the knees and elbows. The ears are rounded and barely reach the nose when pushed forward; the tragus is long and narrow. A prominent keel is present on the calcar. The skull is characterized by a relatively broad interorbital region and a steep forehead with a highly elevated brain-case.

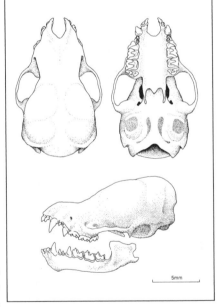

Measurements:
total length: 94 (83-105) n = 33
tail vertebrae: 43 (37-54) n = 33
hind foot: 9 (7-10) n = 31
ear: 12 (9-15) n = 22
tragus: 6 (5-7) n = 17
forearm: 38.3 (34.0-43.0) n = 46
wingspan: 253 (215-272) n = 25
weight: 7.2 (5.5-10.0) n = 18

Dental Formula:
incisors: 2/3
canines: 1/1
premolars: 2/3
molars: 3/3

Identification:
In British Columbia there are three bats with prominently keeled calcars that could be confused with the Long-legged Myotis: the Western Small-footed Myotis (*Myotis ciliolabrum*), the California Myotis (*Myotis californicus*) and the Big Brown Bat (*Eptesicus fuscus*). The Western Small-footed Myotis and California Myotis can be distinguished by their smaller size

(forearm less than 35 mm). The Big Brown Bat has a larger forearm (41-52 mm) than the Long-legged Myotis and lacks fur on its underside from knee to elbow. The skull of the Long-legged Myotis can be discriminated from that of other *Myotis* bats by the relatively broad interorbital area (the ratio of the interorbital to the upper toothrow distances is greater than 0.7 mm) and the strongly elevated brain-case.

Natural History

In British Columbia, the Long-legged Myotis inhabits arid range lands of the interior and humid coastal and montane forests. It ranges from sea level on the coast to 1037 metres elevation in Mount Revelstoke National Park. In the western United States this bat uses buildings, crevices in rock cliffs, fissures in the ground and the bark of trees for summer day roosts. Maternity colonies are situated in attics, fissures in the ground and under the bark of a trees. In southern Alberta, maternity colonies have been found in the crevices of hoodoos. Only two maternity colonies have been found in British Columbia. One was a small colony in the attic of a house on Vancouver Island; the other was a large colony of some 300 individuals that was situated in an old barn near Williams Lake. The only information on summer roosts for males in British Columbia consists of one found in the crack of a dead poplar tree in the Kispiox Valley in July and a juvenile observed among a Yuma Myotis maternity colony in the attic of a church near Squilax in August. Caves and abandoned mine tunnels are exploited for night roosts.

The Long-legged Myotis emerges around dusk and remains active most of the night, even on cool nights—it is relatively tolerant of cold temperatures. This bat is an opportunistic hunter that takes aerial prey over water, forest clearings, among trees and above the forest canopy. Research in Alberta suggests that it prefers to hunt along the edges of tree groves and cliff faces. About 75% of its diet is moths; it also eats termites, spiders, flies, beetles, leafhoppers and lacewings.

There are no winter records of the Long-legged Myotis for the province. In adjacent regions (Alberta, Washington and Montana), it hibernates in caves and mines. At Cadomin and Wapiabi caves in central Alberta, swarming begins in mid August and by late September most individuals are hibernating. Evidently the Long-legged Myotis hibernates in small clusters. No specific information is available on the environmental conditions required for hibernation.

Mating begins in late August or September before the bats enter hibernation. In Alberta, some males breed in their first autumn. The age of sexual maturity for females is unknown; mature females produce a single young. Reproductive data for the province are scanty. Pregnant females have been collected from 23 May to 18 July and nursing females from 25 June to 8 August, suggesting that young are born in late June and July. Recoveries of banded individuals indicate that this mammal can live at least 21 years in the wild.

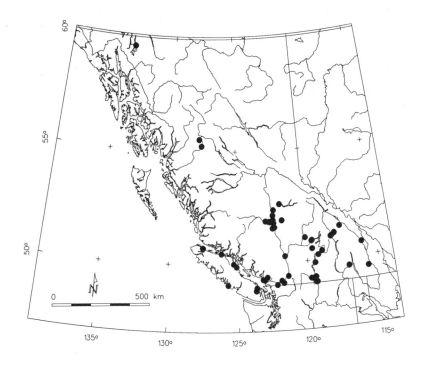

Range

The Long-legged Myotis inhabits western North America from Mexico to southeastern Alaska and western Canada. In coastal British Columbia, it is found on Vancouver Island and the coastal mainland around Vancouver. In the interior, there are records as far north as the Kispiox Valley and Atlin; the eastern limits of the British Columbia range are Cranbrook and Mount Revelstoke National Park.

Taxonomy

All of the British Columbian populations belong to the subspecies *M. v. longicrus*, a race that inhabits the Pacific coast of the United States and western Canada.

Remarks

Although it is one of the more widespread bat species in British Columbia, little is known about the basic biology of the Long-legged Myotis.

Selected References: Dalquest and Ramage (1946), Fenton et al. (1980), Saunders (1989), Schowalter (1980), Warner and Czaplewski (1984).

Yuma Myotis
Myotis yumanensis

Other Common Names: Yuma Bat.

Description

The Yuma Myotis is a medium-size *Myotis*. Its dorsal fur varies from pale brown to nearly black; the fur on its underside is paler. The wing membranes and ears are dark brown. The ears reach the nostrils when pushed forward; the tragus is blunt and about half the length of the ear. The calcar is not keeled. The typical *Myotis* skull has a steeply sloped forehead.

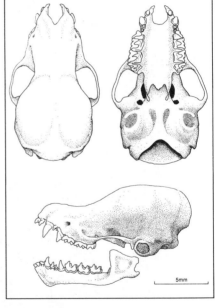

Measurements:
total length: 82 (60-99) n = 322
tail vertebrae: 36 (27-45) n = 324
hind foot: 9 (6-13) n = 323
ear: 14 (8-16) n = 202
tragus: 7 (5-10) n = 212
forearm: 34.3 (30.0-38.0) n = 273
wingspan: 238 (205-260) n = 208
weight: 6.6 (4.0-8.5) n = 153

Dental Formula:
incisors: 2/3
canines: 1/1
premolars: 3/3
molars: 3/3

Identification:
The ear and tragus of the Yuma Myotis are smaller than that of the Fringed Myotis (*Myotis thysanodes*), Northern Long-eared Myotis (*Myotis septentrionalis*), Western Long-eared Myotis (*Myotis evotis*) and Keen's Long-eared Myotis (*Myotis keenii*). The absence of a keel on the calcar distinguishes the Yuma Myotis from the Long-legged Myotis (*Myotis volans*), California Myotis (*Myotis californicus*) and Western Small-footed Myotis (*Myotis ciliolabrum*). This species is most similar in

Yuma Myotis
Myotis yumanensis

external appearance to the Little Brown Myotis (*Myotis lucifugus*)—in some regions it can be extremely difficult to discriminate between the two. For diagnostic external and cranial criteria that identify these two bats see the account for the Little Brown Myotis.

Natural History

This species is restricted to low elevations (sea level to 730 metres) in the province, where it inhabits coastal forests, Ponderosa Pine - Douglas-fir forests and arid grasslands. Its summer day roosts are usually in buildings and other man-made structures in close proximity to water; it has also been found roosting in rock crevices in the Okanagan Valley. Maternity colonies in buildings can be enormous. One of the largest known colonies of bats in British Columbia comprises 1500 to 2000 adult female Yuma Myotis in an old church on the Little Shuswap Indian Reserve near Squilax. Females with young roost in the attic under shingles and boards. Temperatures in this attic may reach 40°C in the midday heat of summer. The degree of clustering in this colony corresponds with temperature: when the air is cool females tend to pack together in close contact, but in the afternoon heat they spread out throughout the attic. The Yuma Myotis also roosts in caves and trees, but colonies in these situations are usually small. Males roost separately from females, either alone or in small groups. (Adult males are virtually unknown in the large colony at Squilax.) Various man-made structures such as house porches, abandoned cabins and bridges serve as night roosts.

Around dusk the Yuma Myotis emerges from its day retreat to hunt over lakes, rivers and streams. Individuals from the Squilax colony travel more than four kilometres to forage over rivers and lakes. In the Okanagan Valley, where this bat mainly feeds over water, aquatic insects such as mayflies, caddisflies and midges are the major prey. Midges are main food in spring; mayflies and caddisflies are the predominant food in summer. Although food habits have not been studied in other parts of the province, aquatic insects are probably the major prey throughout its range given this species tendency to hunt over water. An efficient hunter, the Yuma Myotis can fill its stomach in 10 to 15 minutes on a productive summer night. After feeding, it retreats to a temporary night roost near the feeding area.

Maternity colonies are deserted by late summer or early autumn. In the Squilax colony, the population begins to decline in September and by late October the roost is abandoned. The whereabouts of the Yuma Myotis in winter is unknown. Despite the local abundance of this species, no winter hibernacula have been found in the province. The only known cold-weather record for the province is from Vancouver on 27 March. In coastal Washington a few individuals have been found hibernating in caves—it is possible that similar situations are used as hibernacula in British Columbia.

Mating occurs in autumn and in British Columbia females bear a single young between early June and mid July. Birth dates for the Squilax colony range from 5 June to 21 July with most between 18 June and 7 July. Females generally breed in their first autumn, but the age when males reach sexual maturity is unknown.

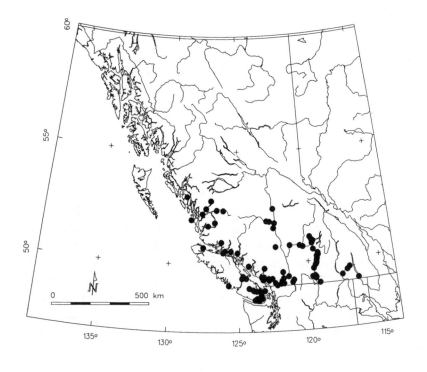

Range

The Yuma Myotis is found across western North America from Mexico to southern British Columbia, where it inhabits several coastal islands including Vancouver Island, the coastal mainland as far north as Kimsquit, and the interior north to the Williams Lake region and east to Nelson.

Taxonomy

There are two races in the province: *M. y. saturatus*, a dark coastal subspecies ranging from California to British Columbia, and *M. y. sociabilis*, a paler race found in the western United States and dry interior of British Columbia.

Remarks

No other bat in the province is so closely associated with water. In many locations it is the most common species captured in mist nets or bat traps set across streams and rivers. The Yuma Myotis is one of the few bats that has been observed flying over salt water in the Pacific Northwest.

There is considerable interest in preserving the large maternity colony at Squilax; this important site is listed with Bat Conservation International (see the Conservation Section in General Biology).

Selected References: Aldridge (1986), Brigham et al. (1992), Dalquest (1947), Herd and Fenton (1983).

Western Red Bat
Lasiurus blossevillii

Western Red Bat
Lasiurus blossevillii

Other Common Names: Southern Red Bat, Red Bat

Description

The Western Red Bat is one of our most distinct bats because of its fur colour, which ranges from pale orange to rusty red (see inside front cover). The dense fur on its back is long and soft. Individual hairs are black at the base, pale yellow in the middle and reddish at the tip. Faint frosting is sometimes evident on the tips of hairs around its neck and back. The bat's underside is paler, varying from yellow to orange. Adult males tend to be brighter in colour than females. There are patches of yellowish fur on the shoulder, bases of thumbs and the underside of the wing membrane. The dorsal surface of the tail membrane and hind feet are thickly furred. The ears are short and rounded with a short, blunt tragus. The hind foot is small; the calcar has a keel. The skull has a short, broad rostrum with broad molars; the brain-case is high and rounded.

Measurements:
total length: 107 (87-120) n = 98
tail vertebrae: 50 (35-60) n = 98
hind foot: 9 (6-13) n = 97
ear: 12 (9-14) n = 81
tragus: 6 (5-7) n = 26
forearm: 38.8 (34.0-42.0) n = 19
wingspan: 278 (270-285) n = 2
weight: 10.8 (7.2-18.5) n = 73

Dental Formula:
incisors: 1/3
canines: 1/1
premolars: 2/2
molars: 3/3
The tiny first upper premolar is usually hidden between the canine and second upper premolar.

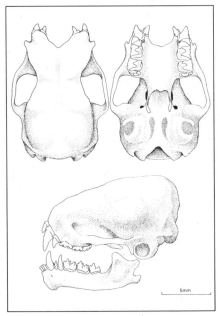

5mm

Identification:

The red fur colour of the Western Red Bat is unique among British Columbian bats. Its skull resembles that of the Hoary Bat (*Lasiurus cinereus*) in shape and dental formula, but it can be distinguished by its smaller size (skull length less than 15 mm).

Natural History*

The scanty information available for British Columbia suggests that this species is restricted to lower elevations where it is associated with forested habitats along rivers. The Western Red Bat is considered to be one of our tree bats. It is a solitary mammal that roosts in the foliage of large shrubs and trees in habitats bordering forests, rivers, cultivated fields and urban areas. Summer roosts, which are occupied by single males or females with young, are located in the branches of deciduous trees. In California the Western Red Bat often roosts in fruit orchards. In the Okanagan Valley, it has been observed flying above stands of cottonwood trees—it is likely that these trees are used for roosting. Red bats tend to roost at lower heights than the Hoary Bat with most sites one to five metres above the ground. Presumably because of the greater protection from predators and disturbance, females with young tend to select higher roosting sites than males.

Hunting begins one or two hours after sunset. Large moths (bodies longer than 10 mm) are the major prey, but the remains of beetles and grasshoppers have been recovered in the guano of the Western Red Bat. Fast flyers, red bats hunt in a straight line or great circles, changing direction suddenly when it detects an insect and attacks. An attack can last as long as five seconds and it may take three or four attacks before the prey is captured. A red bat usually catches its prey in the tail membrane, which can be curled into a pouch, and snatches it up with the mouth. Red bats vary their echolocation calls depending on the situation: long calls for hunting in open areas at long range (five to ten metres) and short calls for hunting in tight situations, such as forest clearings. Red bats seem especially attracted to lights where large numbers of insects are concentrated. Recent studies have revealed that the

* Little research has been done on the Western Red Bat; therefore, we relied extensively on the large body of literature that exists for the Eastern Red Bat for writing this section. Nevertheless, we caution the reader that the degree of similarity in the biology of these two species is unknown at present.

Eastern Red Bat will eavesdrop on the sonar calls of other red bats to locate potential insect prey. There is some evidence that individuals have distinct echolocation calls. This may be important for individual recognition in communication or it may enable red bats to discriminate their own echoes when hunting.

Red bats are migratory, although specific migration movements are poorly documented. No winter records exist for Canada; Canadian populations are assumed to overwinter in southern latitudes. In Canada, the autumn migration period of the Eastern Red Bat extends from late August to October; spring migrants return in May. Evidently some populations of the Eastern Red Bat migrate to southern regions of the United States where they may remain active in winter; other populations overwinter in middle latitudes where most hibernate.

Migratory movements of the Western Red Bat in British Columbia are unknown. However, this species is known to overwinter in the coastal lowlands of California where it roosts in large shrubs and fruit trees. Although the Californian populations are occasionally exposed to freezing temperatures, this species has been observed hunting in midwinter in the San Francisco area. The Eastern Red Bat is a winter resident in a number of eastern states including regions where it would encounter freezing temperatures. In contrast to bats that hibernate in roosts where the temperature is relatively stable, red bats hibernate in trees where they are exposed to great fluctuations in temperature.

Red bats demonstrate several adaptations for hibernating in exposed situations. Their thick fur and small ears assist in minimizing heat loss, and when the temperature falls below 5°C they will cover their ventral region with the tail membrane to conserve heat. When torpid, red bats respond slowly to abrupt temperature changes. Laboratory experiments have demonstrated that individuals will arouse from torpor only when temperatures are raised to 20°C. This enables them to avoid the high energy demands of frequent arousal.

Breeding data for red bats are mostly limited to the eastern species. Mating takes place in flight during migration in late summer or early autumn. Following a gestation period of 80 to 90 days, one to four young are born, with three or four typical of most females. In eastern Canada, young are born in early or mid June. The scanty information available for the Western Red Bat suggests that three young are most common, born in June or early July. Newborns are small, naked and underdeveloped, weighing only 0.5 grams; the eyes are closed. In three to six weeks the young are capable of flight. The female leaves the young at the roost when hunting at night, but she can carry her young when changing roost locations.

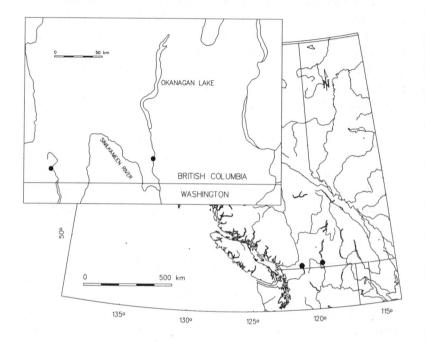

Range

The Western Red Bat is found in Central and South America and the western United States. In British Columbia, it has been found only in the Skagit and Okanagan valleys. With the nearest established populations located in northern California and Utah, these Canadian records presumably represent isolated populations at the northern edge of the summer range.

Taxonomy

It has been long known that red bats of western North America are smaller and more brightly coloured than those in eastern populations of

Lasiurus borealis. In most publications the western form has been treated as a separate race, *L. borealis teliotis*. However, recent genetic research suggests that red bats of western North America and South America warrant recognition as a distinct species, *Lasiurus blossevillii*. According to this taxonomy the British Columbia population is classified in the subspecies *L. blossevillii teliotis*, a race that inhabits Mexico and the western United States.

Remarks

This mammal was first discovered in the province in 1905 when William Spreadborough collected an adult female in July on the Skagit River about 23 kilometres south of Hope. This occurrence was so far north of the known range that Cowan and Guiguet, in The *Mammals of British Columbia*, considered it to be accidental. In 1982, however, Brock Fenton and colleagues detected echolocation calls of this species along the Okanagan River near Okanagan Falls. Their observations suggest that a summer population resides in the Okanagan Valley.

Despite their bright colour, red bats are well concealed when roosting in foliage because they resemble dead leaves.

The Western Red Bat is on the provincial Red List. Based on existing information, it appears to be one of the rarest bats in the province and comprehensive studies should be undertaken to delimit its range and roosting habits. Any observations of this species should be carefully documented and reported to the Royal British Columbia Museum or the Ministry of Environment.

Selected References: Baker et al. (1988), Constantine (1966), Fenton et al. (1983), Grinnell (1918), Hickey and Fenton (1990), Ross (1967), Shump and Shump (1982a).

Hoary Bat
Lasiurus cinereus

Hoary Bat
Lasiurus cinereus

Other Common Names: None.

Description

The Hoary Bat is the largest bat in British Columbia. It has a distinctive hoary colour (see inside back cover). The long, soft fur on its back is a mix of dark brown and grey hairs that are tinged with white. There are small patches of yellow or white on its shoulders and wrists. Yellow fur also appears on the throat, around the ears and on the underside of the wing membranes. The wing membranes are dark brown with paler areas on the forearm and fingers. The ears are round and short with a short, broad tragus. The outer edge of the ear is black; yellow hairs are scattered on the inside of the ear. The dorsal surface of the tail membrane is densely furred; the hind foot is relatively small with a heavy covering of fur on the upper surface. The calcar has a narrow keel. The skull has a short, broad rostrum, large molars and a high profile.

Measurements:

total length: 137 (125-144) n = 11
tail vertebrae: 60 (50-66) n = 10
hind foot: 12 (10-15) n = 12
ear: 14 (13-16) n = 4
tragus: 9 (9-10) n = 4
forearm: 54.5 (50.3-57.4) n = 24
wingspan: 392 (338-415) n = 7
weight: 28.4, (20.1-37.9) n = 43

Dental Formula:

incisors: 1/3
canines: 1/1
premolars: 2/2
molars: 3/3

The tiny first upper premolar is usually hidden between the canine and second upper premolar.

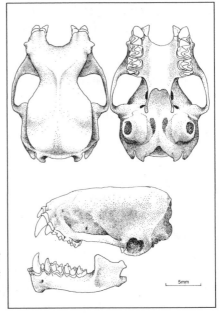

5mm

Identification:

The large size and marked colour pattern make the Hoary Bat one of our most distinct bats. The only species with remotely similar markings is the Silver-haired Bat (*Lasionycteris noctivagans*)—see that account for diagnostic external features. Among British Columbian bats, only the Western Red Bat (*Lasiurus blossevillii*) has a similar dental formula; the skull of the Hoary Bat can be discriminated by its greater size (skull length greater than 15 mm).

Natural History

The Hoary Bat is associated with a variety of forested and grassland habitats in the province. Its elevational range is from sea level to 1250 metres.

Because of its tendency to roost in the branches of coniferous and deciduous trees, the Hoary Bat is often referred to as a tree bat. Specific details on tree roosts in British Columbia are not available—only a few individuals have been found roosting in the branches of fruit trees in the Okanagan. In Manitoba the Hoary Bat roosts some 8 to 12 metres above the ground, usually near the ends of branches of deciduous trees such as Green Ash. It seems to select sites that will conceal it from predators and yet also provide an open flight path for easy access to and from the roost. In Oregon the Hoary Bat prefers old Douglas-fir forests, presumably because of its tree-roosting habits. Although its usual day roosts are in the branches of trees, the Hoary Bat has been found roosting in tree cavities in British Columbia: one was collected from a hollow cedar tree in Garibaldi Provincial Park and another was found in a woodpecker nest in a tree cavity. The Hoary Bat rarely roosts in caves or buildings.

In summer, males are solitary and females roost with their young. Unlike most of our bats, females do not congregate in maternity colonies. Family groups will use the same roost for more than a month.

Hoary Bats emerge about 30 minutes after sunset to feed and continue to forage throughout the night. Nursing females reduce their foraging time during the the first few weeks of nursing to spend more time with their young at the roost. Because their tree roosts are more exposed than the maternity roosts of other bats, females will stay with their undeveloped young for longer periods to keep them warm.

The Hoary Bat is suited to preying on large insects, with its large skull and teeth and its swift flight. It emits low-frequency echolocation calls that are most effective for detecting large prey at long range. The Hoary Bat hunts at tree-top level in open areas such as fields and forest clearings. Large moths, beetles and dragonflies form the bulk of its diet; small insects, such as midges and flies, are less common prey.

The Hoary Bat is often attracted to insect concentrations at lights outside buildings; permanent outdoor lights may even be responsible for its presence at some locations in the province. In well-lit areas where insect prey are concentrated, a Hoary Bat will establish a feeding area and chase away other bats (including other Hoaries). Loud chirping calls audible to the human ear are often emitted during these chases. Besides their role in communication, these calls may also provide some echolocation information.

Circumstantial evidence based on seasonal occurrences in British Columbia strongly suggests that the Hoary Bat is migratory. There are records from 19 June to 15 October with most from mid August to early October, the period when the autumn migration is presumed to take place. The winter range of British Columbian populations is unknown. It appears that populations in the Pacific Northwest migrate to southern California or Mexico for the winter.

Mating likely takes place during the autumn migration or in winter; females are pregnant during the spring migration. Although there are no breeding data available for British Columbia, in other parts of Canada young are born in June and nurse from early June to the end of July. Normally females leave the young at the roost at night, but they can carry their young for about a week after birth. Females typically give birth to twins, although one, three or four young have also been observed. The newborn bats are undeveloped with closed eyes and ears. The top of the head, shoulders and tail membrane are covered with fine, silver-grey hairs and the underside is naked. Development is slow with the ears opening within three days and the eyes in around twelve days. By five weeks the young are capable of sustained flight. Even after the young are flying, family groups remain together for several more weeks. The relatively lengthy period of parental care shown by female Hoary Bats may be because this mammal does not hibernate—it has fewer energy demands than hibernating bats, which have to accumulate fat reserves for winter.

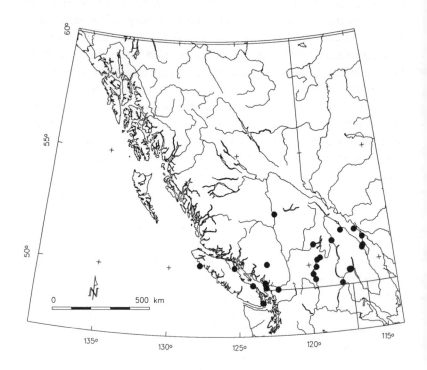

Range

The Hoary Bat has the broadest distribution of any North American bat ranging from South America to northern Canada. It has managed to colonize a number of isolated islands including the Hawaiian Islands. In British Columbia, it is found on Vancouver Island, the coastal mainland north to Garibaldi Provincial Park, and the southern interior north to the Williams Lake region. In some regions of North America the sexes appear to occupy separate summer ranges. Based on the few available records the sexes appear to overlap extensively in British Columbia.

Taxonomy

One subspecies, *L. c. cinereus*, is recognized in North America.

Remarks

The thick, heavy fur covering on the body and tail membrane of this bat is important for insulation. The hoary colour provides camouflage against a background of lichen-covered bark. Because of its solitary roosting habits in trees, the Hoary Bat is rarely encountered by man. There is still much to be learned about its distribution and roosting behaviour in the province. It is a bat that can be easily identified and we encourage naturalists to report any observations.

Selected References: Barclay (1984), Barclay (1985), Barclay (1986), Barclay (1989), Findley and Jones (1964), Perkins and Cross (1988), Shump and Shump (1982*b*).

Silver-haired Bat
Lasionycteris noctivagans

Silver-haired Bat
Lasionycteris noctivagans

Other Common Names: None.

Description
The fur of the Silver-haired Bat is dark brown or black with scattered silver-white-tipped hairs giving it a lightly-frosted appearance. Old individuals tend to have fewer white-tipped hairs and their fur often appears pale brown or yellowish. On the underside, the frosted hairs are concentrated in the belly. The ear is short and round with a short, blunt tragus. The ears and wing membranes are black. The dorsal surface of the tail membrane is lightly furred. The calcar lacks a keel. The skull has a blunt rostrum and is flat in profile.

Measurements:
   total length: 100 (90-117) n = 68
   tail vertebrae: 41 (31-50) n = 70
   hind foot: 9 (6-11) n = 69
   ear: 12 (9-15) n = 17
   tragus: 7 (4-8) n = 22
   forearm: 41.4 (39.1-43.9) n = 56
   wingspan: 291 (200-354) n = 38
   weight: 9.0 (5.8-12.4) n = 14

Dental Formula:
   incisors: 2/3
   canines: 1/1
   premolars: 2/3
   molars: 3/3

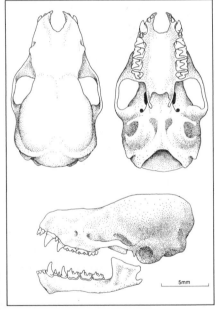

Identification:
The only Canadian species this bat could be confused with is the Hoary Bat (*Lasiurus cinereus*). But the Silver-haired Bat is much smaller (forearm less than 44 mm). Also its light covering of fur at the base of the tail membrane contrasts with the thickly furred dorsal surface of the Hoary Bat's tail membrane. Among British Columbian bats, only Townsend's Big-

eared Bat (*Plecotus townsendii*) has the same dental formula. The skull of the Silver-haired Bat can be readily distinguished from that of Townsend's Big-eared Bat by its flat profile.

Natural History

In British Columbia, the Silver-haired Bat is associated with forest and grassland habitats. The elevational range extends from sea level to 1220 metres. This species is generally regarded as a tree bat, although specific information on its summer roosting habits is rather sketchy. Individuals have been found under the bark of trees and in crevices in tree trunks, abandoned woodpecker holes and bird nests. In the Okanagan Valley cottonwood trees appear to be important roosting sites.

Typically the Silver-haired Bat roosts alone or in small groups of two to six. Even when roosting in groups, individuals are rarely found in close contact. Only two maternity colonies have been found in Canada. Both were small—one with three and the other with eight females—and were situated in cavities excavated by woodpeckers.

This species emerges about 30 minutes after sunset to hunt in clearings around tree-top level and over water. There are two well-defined peaks in activity: the first between 10 pm and midnight, and the second an hour before dawn. Night-time activity is sharply curtailed when temperatures are below 8°C. Its slow, agile flight and high-frequency echolocation calls enable the Silver-haired Bat to readily detect and intercept small insects at close range. It seems to be particularly adept at exploiting swarms of flying insects. Although some have labelled the Silver-haired Bat a moth specialist, its diet is extremely flexible—it appears to be able exploit whatever insect prey is available. Prey items identified in the diet cover a wide range of small insect species including moths, midges, leafhoppers, caddisflies, flies, beetles, ants and termites. No information is available on this bat's use of night roosts.

Most Canadian populations of the Silver-haired Bat are thought to be migratory, overwintering in the United States. Changes in seasonal occurrence suggest that its range shifts northward in spring and southward in winter. But in British Columbia this bat can be found in all seasons. Given this evidence and considerable overlap in the summer and winter ranges, some researchers have concluded that British Columbian populations of the Silver-haired Bat do not migrate. But the majority of records from the province come from spring (May and

June) and late summer (August and September), periods when this bat is presumed to migrate. With no information available on movements of individuals, we cannot determine whether these seasonal peaks in occurrence result from an increase in bat activity or an influx of migrating individuals.

During the migration period the Silver-haired Bat uses several different kinds of temporary day roosts. In Manitoba it has been found roosting in the furrowed bark and crevices of large ash and willow trees. Presumably, these crevice roosts provide protection from inclement weather and predators. In urban areas, migrating Silver-haired Bats often roost on the sides of buildings. Each autumn the Royal British Columbia Museum receives several reports of this species roosting on buildings in downtown Victoria. During the day individuals can lower their body temperature and go into torpor in order to conserve energy. Individuals found in this dormant condition are often assumed to be ill. Sometimes during migration periods, this bat is caught in mist nets set by bird banders.

There are winter records from Victoria and Vancouver and from the Okanagan Valley north to Williams Lake. Trees appear to be the most important hibernation sites in the province: Silver-haired Bats have been found hibernating under the bark of Western Red-cedars in Vancouver; one was discovered in a Douglas-fir snag near Kamloops. There is one winter record of an individual hibernating in the attic of a house in Vancouver. Also an active individual was found in an old mine in the south Okanagan in January. Other than it being a cold-hardy bat, little seems to be known about the Silver-haired Bat's specific requirements for hibernation. In old mines, it has been found hibernating in ambient temperatures of -0.5 to -2°C. The northern limits of the winter range suggest that this mammal can overwinter in regions where the average daily temperature for January is above -7°C.

Mating is thought to take place during the autumn migration period. Most males and females reach sexual maturity in their first summer. After a gestation period of 50 to 60 days, females give birth to one or two young, with twins more common. The only breeding datum for the province consists of a female collected in the Peace River region on 18 June with two foetuses seven millimetres in size. Birth probably occurs in late June or early July. Newborn young are hairless and weigh about two grams. They develop quickly and by three weeks they can fly.

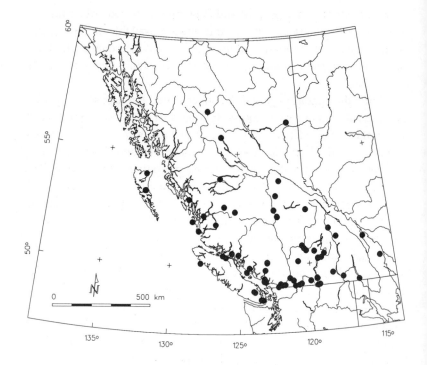

Range

This bat ranges widely in North America from northernmost Mexico to southern Canada and southeastern Alaska. In British Columbia it inhabits several coastal islands including Vancouver Island and the Queen Charlotte Islands, the coastal mainland north to Rivers Inlet, and the interior as far north as the Peace River and Spatsizi Plateau. In some parts of North America, males and breeding females appear to occupy separate summer ranges. This may not be the case in British Columbia because male and female distributions overlap extensively in all seasons. Although there is some speculation that the sexes could occupy different elevations in summer, there are too few records to evaluate the elevational range of the sexes in the province.

Taxonomy

No subspecies are recognized.

Remarks

Because the Silver-haired Bat uses trees for day roosts, maternity colonies and hibernacula it is a species that may be especially vulnerable to deforestation and the removal of snags. The impact of forestry practices on this bat needs to be assessed. Donald Thomas's research in the forests of coastal Oregon and Washington revealed that Silver-haired Bat activity was greatest in stands of old-growth forest; he attributed this to the availability of abundant roosting sites that are associated with snags and old trees.

Because it is a difficult species to find and study, there is still a great deal to be learned about the seasonal distribution and movements of the Silver-haired Bat in British Columbia. This bat can be easily identified in the hand or by close observation and we encourage naturalists to record any observations.

Selected References: Barclay (1985), Barclay (1986), Barclay et al. (1988), Cowan (1933), Kunz (1982), Parsons et al. (1986), Schowalter et al. (1978).

Big Brown Bat
Eptesicus fuscus

Big Brown Bat
Eptesicus fuscus

Other Common Names: None.

Description

One of our larger bats, the Big Brown Bat has a large, broad head, broad nose and long, lax fur. Its fur colour varies from pale to dark brown; the fur tends to be oily in texture. Its flight membranes and ears are black. The ears just reach the nose when pushed forward; the tragus is short and blunt. The calcar has a prominent keel. The skull is robust, with thick, heavy jaws, a flattened brain-case and large teeth.

Measurements:

total length: 116 (98-131) n = 126
tail vertebrae: 46 (37-55) n = 123
hind foot: 12 (8-15) n = 115
ear: 15 (11-21) n = 53
tragus: 8 (5-11) n = 40
forearm: 47.5 (43.0-52.0) n = 109
wingspan: 328 (205-393) n = 62
weight: 15.2 (8.8-21.9) n = 36

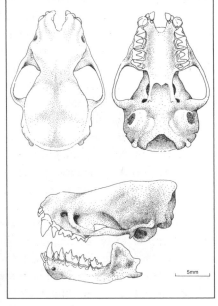

Dental Formula:

incisors: 2/3
canines: 1/1
premolars: 1/2
molars: 3/3

Identification:

Although the Big Brown Bat resembles several of our *Myotis* species in colour, it can be readily distinguished by its large size (forearm greater than 42 mm), large head and long fur. The combination of 2/3 incisors and 1/2 premolars is unique among British Columbian bats.

Natural History

The Big Brown Bat inhabits a variety of habitats in the province, including arid grassland and interior and coastal forests. Its elevational range is from sea level to 1070 metres. Throughout its range this bat demonstrates a strong affinity for buildings and it is often referred to as a house bat or barn bat. In British Columbia, maternity colonies have been found in the attics of houses, cabins and barns. But in the Okanagan Valley this species rarely roosts in man-made structures; there, maternity colonies have been found in cavities in dead Ponderosa Pines and rock crevices.

The Big Brown Bat is less tolerant of high temperatures than other species that use attics such as the Little Brown Myotis and Yuma Myotis. Temperatures above 35°C will often force the Big Brown Bat to move to a cooler site within the roost or change roosts altogether. Maternity colonies as large as 700 individuals have been reported, but those associated with buildings in British Columbia are small, comprising no more than 50. The tree colonies studied in the Okanagan contain any number from a few to 200 bats, with an average of about 100. Loyalty to roosting sites seems to depend on the type of roost. Big Brown Bats roosting in buildings or rock crevices show a strong fidelity to those sites; however, those roosting in trees seem to move regularly between several established roosts. Some recent genetic research has revealed that most individuals in maternity colonies are closely related.

This species emerges around dusk to feed. It is regarded as a generalist that hunts in a variety of situations: over water, over forest canopies, along roads, in clearings, and in urban areas, often around street lights that attract insects. The intensity of its echolocation calls is very high. In open areas such as forest clearings and over water the Big Brown Bat may first detect prey as far as 5 metres away. In most locations beetles are an important part of the diet and the heavy teeth and strong jaws of this bat appear to be an adaptation for chewing hard-shelled insects. Nevertheless, remains of moths, termites, carpenter ants, lacewings and various flies also have been identified in faecal pellets.

The only data on feeding biology for the province comes from research undertaken at Okanagan Falls. There the Big Brown Bat emerges 30 to 40 minutes after sunset and hunts 5 to 10 metres above the water along a 300-metre stretch of river. It is a relatively fast flyer, able to attain a speed of 4 metres per second when pursuing prey. The Big Brown Bat makes several feeding flights during the night, each 30 to 60 minutes in duration. Between flights it usually roosts near the feeding area. During the day it roosts as far as 4 kilometres from the feeding area, even though there are sites near the Okanagan River suitable for day roosts.

Large caddisflies are the major prey at Okanagan Falls. Midges are the dominant insect group in the area, but the Big Brown Bat does not eat them. The echolocation calls of this bat should enable it to detect midges at a distance of about a metre. Either it is unable to react quickly enough to capture them or it simply prefers larger prey.

This species' hibernation period varies with local climate and geography. In the interior of the province, it lasts from November to April. In British Columbia's southern regions, the Big Brown Bat may be active for short periods in winter. Migrations of almost 300 kilometres have been documented, but most individuals travel no more than 80 kilometres between summer and winter roosts. Populations exploiting buildings may even use the same site in winter and summer. (Despite its rather sedentary nature, experiments have demonstrated that this species possesses a remarkable homing ability: individuals released 400 kilometres from their roosts managed to find their way back.)

The Big Brown Bat hibernates in buildings, caves and mines; in western Canada most populations appear to use buildings. Hibernating colonies are typically small, although large colonies have been found in a few caves and mines in eastern North America. As with most of our bats, information on winter biology in British Columbia is sketchy and derived largely from incidental information on museum specimens. There are coastal winter records from Victoria and interior records from as far north as Williams Lake and Prince George. Most of these winter records are from buildings; this mammal is rarely found in caves or mines in the province.

In eastern Canada, swarming behaviour often takes place at mines and caves before this species enters hibernation. The Big Brown Bat hibernates alone or in clusters of as many as 100. This is a hardy bat capable of withstanding low temperatures. It selects a dry microenvironment with temperatures of -10 to -5°C, although temperatures below -4°C will usually stimulate arousal. When the Big Brown Bat enters hibernation in late autumn it weighs about 25 grams; when it completes hibernation in spring it weighs about 19 grams, having lost about 25% of its body weight. In caves or mines it hibernates in exposed areas near the entrance where it hangs from the ceiling and walls, or it wedges itself into small crevices or under rocks. Big Brown Bats often arouse from hibernation to change roosting locations; there are several observations of this bat outside on the walls of buildings when temperatures were well below freezing. There is some evidence that winter activity is more common in buildings where conditions for hibernation may not be ideal.

Mating is thought to take place in autumn and winter. Fertilization occurs in spring after departure from the hibernaculum. Males may attain sexual maturity in their first year; females may breed in their first year but many do not breed until their second. Although twins are typical in eastern North America, single young are usually produced in the west. The scanty data available for the province suggest that the young are born in late June. Nevertheless, birth dates and development of the young can be extremely variable, differing among roosts and even among individuals within a roost. Pregnant females, females with newborn young and females with well-developed young can be found together in the same roost. In the Okanagan, considerable year-to-year variation has been observed, with dates for the first appearance of the young varying as much as a month. This variation probably results from year to year differences in climate. Newborn young weigh about three grams. They are naked and blind although the eyes open within a few hours after birth. Juveniles are capable of flying at 18 to 35 days of age. There are longevity records showing that this species can live at least 19 years in the wild.

Range

The Big Brown Bat has a vast range that extends from northern South America to southern Canada. In British Columbia, it is found on Vancouver Island, the coastal mainland north to the Bella Coola River Valley and the interior where its northern limits are unknown. Northernmost localities in the province are from the Prince George and the Peace River region, but because there is a record from the interior of Alaska the range may extend into extreme northern British Columbia.

Taxonomy

All British Columbian populations belong to the race *E. fuscus bernardinus*, a subspecies that ranges from California to British Columbia.

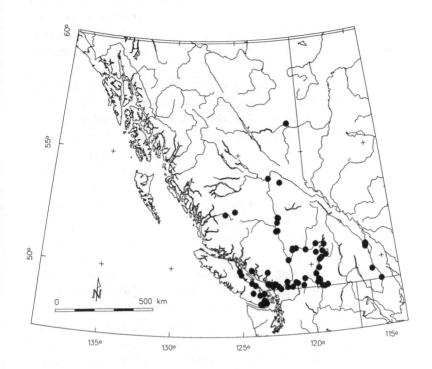

Remarks

This species may have some claim to being the "provincial bat" because it often turns up in the provincial legislative buildings in Victoria during autumn and winter.

Selected References: Brigham (1990), Brigham (1991), Brigham and Fenton (1986), Brigham and Fenton (1991), Kurta and Baker (1990), Schowalter and Gunson (1979).

Spotted Bat
Euderma maculatum

Spotted Bat
Euderma maculatum

Other Common Names: Pinto Bat.

Description

Jet black dorsal fur with large white spots on the rump and shoulders make the Spotted Bat one of the most distinct and spectacular bats in North America (see inside back cover). It has smaller white patches at the base of each ear and its underside is whitish with black underfur. There is a small, naked patch on its throat, which is often hidden in the fur. The immense pinkish ears are joined at their bases across the forehead; the tragus is large and broad. A fringe of fine hairs extends along the top border on the back of the ears. The calcar is not keeled. The skull has an elongated brain-case with a gradually sloping forehead.

Measurements:

total length: 116 (107-125) n = 4
tail vertebrae: 50 (47-50) n = 4
hind foot: 10 (9-10) n = 4
ear: 38 (34-41) n = 5
tragus: 13 (13-14) n = 3
forearm: 50.9 (47.9-53.1) n = 9
wingspan: 346 (336-355) n = 5
weight: 17.9 (16.2-21.4) n = 6

Dental Formula:

incisors: 2/3
canines: 1/1
premolars: 2/2
molars: 3/3

Identification:

The conspicuous ears and three white spots on the back distinguish this species from any other British Columbian bat. The combination of 2/3 incisors and 2/2 premolars is also unique among British Columbian bats.

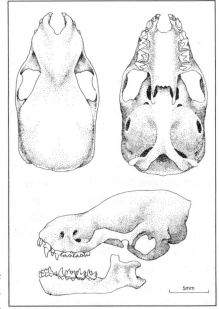

Natural History

Throughout western North America the Spotted Bat is associated with arid desert terrain. Typically, day roosts are in crevices of steep cliff faces. The most important roosting sites in the Okanagan Valley are located in cliffs at McIntyre Bluff, Gallagher Bluff, Spotted Bluff, the west side of Vaseux Lake and the Vaseux canyon. The Spotted Bat hunts over open Ponderosa Pine forests, hay fields and marshes adjacent to lakes. Every summer it can be found foraging over the marshy area on the west side of Vaseux Lake. Its elevational range in British Columbia extends from 300 to 900 metres above sea level, although most occurrences are below 500 metres.

The Spotted Bat leaves its roost 30 to 60 minutes after sunset. It is a solitary hunter that feeds on moths captured while flying 5 to 15 metres above the ground. Although large ears are often associated with gleaning bats, there is no evidence to suggest that the Spotted Bat is a gleaner. It is quite predictable in its daily movements, usually following a set route to its night-time feeding area and returning to the same roost night after night. It may move as far as 10 kilometres between the day roost and feeding areas.

Its echolocation calls range between 6 and 16 kilohertz, making this our only bat that can be heard without a bat detector, at least by those whose ears are relatively sensitive to high-frequency sounds. The calls resemble a high-pitched, metallic click and could be mistaken for insect sounds by those unfamilar with them. The calls of the Spotted Bat have several important characteristics that affect its natural history. Low-frequency calls should be most effective for detecting larger prey, which is consistent with the evidence suggesting that most of this bat's diet consists of medium-sized moths. Several researchers have suggested that the low-frequency calls enable the Spotted Bat to avoid detection by certain species of moths that can detect the ultrasonic echolocation calls of other bats. Another characteristic of low-frequency sounds is that they can carry for considerable distances through the air—it is suspected that the Spotted Bat can detect its insect prey at long range. The extraordinarily large ears probably help the bat hear the returning echoes of its low-frequency calls.

The audible echolocation calls make it relatively easy for the biologist or naturalist to monitor the presence and activity of this bat. Even so, the distribution of Spotted Bats in the province is poorly known. The recent discovery of this species in the Thompson, Fraser and Chilcotin valleys indicates that there may be localized populations throughout the dry interior. We strongly encourage local naturalists in this region to familarize themselves with the sounds of this bat and to search for it in their area.

The Spotted Bat has been heard in the Okanagan from early April to late October, but the whereabouts of this population in winter is a mystery. Little is known about the winter biology of this bat throughout its entire range. The only hibernation record for its entire range is of four individuals in a cave in Utah some sixty years ago. Research in southern Utah suggests that this species is occasionally active during mid winter in the western United States. No hibernating individuals have been found in British Columbia. The Spotted Bat may hibernate in crevices in cliffs inaccessible to humans, or it may spend winter outside the province.

Little information is available on reproduction but the scanty data suggest that mating takes place in autumn. Females produce one young; the only pregnancy record for British Columbia suggests that young are born in late June or early July.

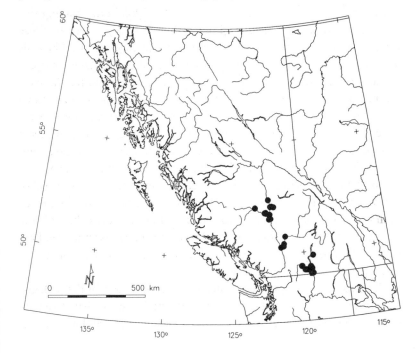

Range

This is a rare bat with a discontinuous distribution throughout its range across western North America. Its range includes Mexico, and the United States from Arizona and California to Idaho and Montana. In Canada, the Spotted Bat occurs only in British Columbia where there are localized populations in the dry interior from the southern Okanagan Valley to the Chilcotin River and Williams Lake region.

Taxonomy

No subspecies are recognized.

Remarks

The Spotted Bat is on the provincial Blue List. This mammal was first discovered in British Columbia in 1979 when Brock Fenton and his colleagues detected its echolocation calls near Oliver while conducting a general bat survey in the Okanagan.

The pattern of white markings on the Spotted Bat resembles the patterns found on some moths. The significance of these markings is unknown. They could serve as a warning signal to predators, provide camouflage or act as some kind of visual signal in communication, but there is no evidence to support any of these ideas.

Selected References: Leonard and Fenton (1983), Poché (1981), Wai-ping and Fenton (1989), Watkins (1977), Woodsworth et al. (1981).

Townsend's Big-eared Bat
Plecotus townsendii

Other Common Names: Lump-nosed Bat, Western Big-eared Bat.

Description

Townsend's Big-eared Bat is a medium-size bat with enormous ears—about one half its body length—and two prominent, glandular swellings on its nose. Its long dorsal fur varies from pale brown to blackish-grey; hairs in the underfur are paler. The tragus is long and pointed—about one third the ear length. The calcar lacks a keel. The skull is relatively narrow and the profile of the brain-case is curved.

Measurements:

total length: 100 (83-113) n = 52
tail vertebrae: 46 (38-57) n = 44
hind foot: 11 (7-10) n = 51
ear: 34 (27-40) n = 26
tragus: 13 (10-15) n = 11
forearm: 42.6 (39.0-45.2) n = 55
wingspan: 287 (232-313) n = 32
weight: 8.6 (6.0-13.5) n = 15

Dental Formula:

incisors: 2/3
canines: 1/1
premolars: 2/3
molars: 3/3

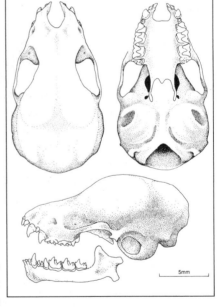

Identification:

The Pallid Bat (*Antrozous pallidus*) and the Spotted Bat (*Euderma maculatum*) are the only other species in the province with such large ears. The Spotted Bat is readily identified by its distinctive markings and the Pallid Bat is larger (forearm length 45-60 mm); neither have the lumps on the nose. Among British Columbian bats, only the Silver-haired Bat (*Lasionycteris noctivagans*) has the same dental formula. The flat profile of the skull distinguishes it from Townsend's Big-eared Bat.

Townsend's Big-eared Bat
Plecotus townsendii

Natural History

In British Columbia this species is associated with a variety of habitats from coastal forests to arid grasslands of the interior. Its elevational range in the province is from sea level to 1070 metres, although most occurrences are from low elevations.

In the western United States, Townsend's Big-eared Bat uses caves, old mines and buildings as summer day roosts, with buildings being used more often in humid coastal areas. It uses similar situations for night roosts. Although Leo Jobin reported this species using old mine shafts near Williams Lake in summer, other researchers did not find it in any caves or mines during recent summer bat inventories in British Columbia.

Females form colonies of a dozen to several hundred in dimly lit areas in buildings, caves or mines. The only nursery colony found in British Columbia was in the attic of a house on Vancouver Island; it consisted of about 60 females and their young. Females and their young form tightly packed clusters that prevent heat loss and ensure the rapid development of the young. This species is particularly sensitive to human disturbance and a number of biologists have noted that females will permanently abandon a traditional summer roost if disturbed. Males roost alone during summer, separate from females.

A late flyer, Townsend's Big-eared Bat emerges an hour or so after dark. It is an agile bat that is capable of flying at slow speeds. Food habits have not been studied in British Columbia. In the western United States, small moths (body length, 3-10 mm) form most of the diet. It also eats lacewings, dung beetles, flies and sawflies. This species feeds several times during the night—it is often near dawn before it returns to the day roost.

In August, nursery colonies break up and individuals begin to migrate to caves and mine adits for hibernation. Townsend's Big-eared Bat is relatively sedentary, moving 10 to 65 kilometres between the summer roost and the winter hibernaculum. This is one of the few bats that has been consistently found hibernating in British Columbia. On the coast, there is a hibernaculum in a cave on Thetis Island that supports a population of 20 to 40 Townsend's Big-eared Bats each winter. The temperature in the cave is usually 8-10°C. In the interior, small hibernating colonies (up to 16 individuals) have been found in mines and caves from the Okanagan Valley to the Williams Lake region. In these interior hibernacula, this species usually occupies dry, exposed locations near the mine entrances where the temperature are 5-8°C. But, in a survey of potential hibernacula conducted in the central interior in February 1990, several Townsend's Big-eared Bats were discovered hibernating in an ambient temperature of -4°C in an old mine shaft on the north shore of Kamloops Lake and a single bat was hibernating in an exposed tunnel of a limestone cave near Williams Lake where the temperature was -7°C. These observations were made a few days after a severe cold spell; they indicate that populations living at the northern limits of the range may tolerate periods of freezing temperatures.

Torpid Townsend's Big-eared Bats have been found in caves or mines from 16 September to 23 May. When in a state of deep torpor this bat hangs from a horizontal surface by its feet; its ears are curled back along the head in a shape resembling a ram's horn. Folding the ears may reduce heat loss. Curiously, the tragus remains erect when the ears are folded—it has been suggested that the tragus acts as a heat sensor. By the end of hibernation, Townsend's Big-eared Bat may have lost more than half its autumn weight. This species frequently arouses from hibernation in response to temperature changes or disturbance—it will change its position within a hibernaculum or even move to another cave or mine site in mid winter. These movements contribute to the severe weight loss, and they place a heavy demand on this bat's limited energy reserves.

Mating takes place from November to February, usually at the winter roost. Males do not breed in their first autumn; but yearling females can breed in their first year, usually giving birth later in the season than older females. A single young is born after a pregnancy of 50 to 100 days. The gestation period is controlled largely by temperature. Cool temperatures will induce torpor in pregnant females and slow down the development of the foetus. In coastal areas the young are probably born in June. When the Vancouver Island nursery colony was examined on 7 July it contained young at various stages of development: some were still nursing and others were capable of flying. The few breeding data available for the interior are limited to the southern Okanagan Valley. A pregnant female was captured on 3 July and a nursing female on 24 July, suggesting that young are born in mid July. At birth, the young weigh about 2.4 grams, their eyes are closed and the ears are not erect.

Young Townsend's Big-eared Bats mature quickly; by three weeks they are capable of flying and at four weeks they are nearly adult size.

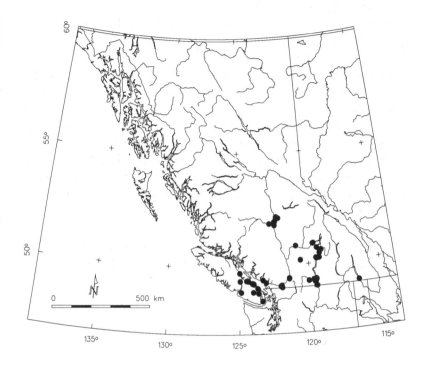

Range

Townsend's Big-eared Bat is found throughout the western United States; there are isolated populations on the southern Great Plains and in the Ozarks and Appalachians. In Canada, it is restricted to British Columbia. On the coast, it inhabits Vancouver Island, the Gulf Islands and the Vancouver area; in the interior, it has been found as far north as Williams Lake and east to Creston.

Taxonomy

Two subspecies occur in the province: *P. t. townsendii*, a dark coastal race that ranges from California to southwestern British Columbia, and *P. t. pallescens*, a paler interior race that inhabits the western United States and the southern interior of British Columbia.

Remarks

Most biologists regard the characteristic large ears of Townsend's Big-eared Bat as an adaptation to its quiet echolocation calls. But some have speculated that the large ears may also provide some lift during flight and assist with temperature regulation by releasing heat. The distinctive glandular lumps on the nose may function as sexual scent glands.

Remains of a Townsend's Big-eared Bat were recovered in the stomach of a Marten trapped on the Klanawa River, Vancouver Island, in mid January.

The provincial Ministry of Environment has listed Townsend's Big-eared Bat on its Blue List. Although it is widespread across most of southern British Columbia, this bat is particularly vulnerable to human activity. Disturbing females with young will affect breeding success, and repeated disturbance at winter hibernacula will increase winter mortality.

Selected References: Graham (1966), Humphrey and Kunz (1976), Jobin (1952), Kunz and Martin (1982), Pearson et al. (1952).

Pallid Bat
Antrozous pallidus

Other Common Names: None.

Description

The Pallid Bat is one of our largest bats; only the Hoary Bat is greater in size. The Pallid Bat is pale, as its name suggests, with short, fine fur. The dorsal fur is pale yellow with a tinge of brown, and the fur on the underside is creamy white. Individual hairs are pale at the base with darker tips. The pale ears are broad and extend well beyond the nose when pushed forward; the tragus is long and narrow, with a toothed outer edge. Compared with other Canadian bats the eyes are relatively large. The snout is square with a shallow ridge on top; the end of the snout is scroll-shaped. The calcar lacks a keel. The skull is large with heavy, robust teeth.

Measurements:
total length: 114 (102-135) n = 50
tail vertebrae: 47 (38-55) n = 50
hind foot: 12 (9-17) n = 51
ear: 29 (26-33) n = 40
tragus: 14 (12-17) n = 34
forearm: 52.5 (48.0-57.4) n = 47
wingspan: 351 (310-370) n = 12
weight: 17.0 (12.0-24.3) n = 29

Dental Formula:
incisors: 1/2
canines: 1/1
premolars: 1/2
molars: 3/3

Identification:
The only species the Pallid Bat could be confused with is Townsend's Big-eared Bat

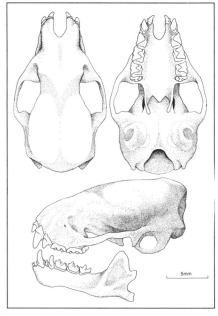

(*Plecotus townsendii*); but the Pallid Bat is much larger, its ears are not joined at their bases and it lacks the two prominent bumps on the nose.

Pallid Bat

Antrozous pallidus

The Pallid Bat is the only British Columbian bat with two pairs of lower incisors.

Natural History

The Pallid Bat is found in arid desert habitat, often near rocky outcrops and water. In British Columbia it is restricted to low elevations (300-490 metres) in grassland areas and Ponderosa Pine forests in the vicinity of cliff faces.

Little is known about its biology in the province, but it has been the subject of intensive study in the southwestern United States. Horizontal rock crevices with a hot, constant temperature (30°C) are the preferred summer day roosts, although it has also been found roosting in tree cavities, buildings, caves, mine adits and crevices in cliffs. A gregarious bat, most of the Pallid Bat's summer roosts comprise 20 to 200 individuals—the largest are maternity colonies. In some areas, males and females roost separately, but in others, mixed colonies containing both sexes have been found. The composition of colonies in British Columbia is unknown, but 14 of 19 Pallid Bats captured in the province were males, suggesting that the population may be predominately male. It appears that Pallid Bats frequently change the location of their day roost.

In summer this species is a late feeder leaving its day roost about 45 minutes after sunset. Pallid Bats may commute up to four kilometres between their day roosts and foraging areas. In British Columbia, this species seems to hunt mainly over tracts of open grassland sparsely covered with Big Sagebrush, Rabbit Brush and Bitter-brush. Gravel roads may provide foraging corridors—there are several observations of Pallid Bats flying low over roads in the Okanagan.

A large, powerful bat with robust teeth, this species is well adapted for killing and eating large invertebrates. It usually gleans prey from the ground and the foliage of trees and shrubs, but it occasionally pursues insects in the air. In the western United States the Pallid Bat eats june beetles, moths, cicadas, praying mantises, katydids, grasshoppers, scorpions and crickets. There are also records of this bat preying on small lizards and even a desert Pocket Mouse. Faecal pellets from the Okanagan Valley contained mostly beetles; moths and lacewings were minor prey items. While hunting, the Pallid Bat flies slowly, close to the ground, with rhythmic dips and rises. While hunting prey on the ground it listens for their rustling sounds; it will also forage in the foliage of trees and shrubs. Aerial prey are tracked with echolocation. The Pallid Bat eats small prey while it is flying, but takes larger items back to its night roost for consumption.

Night roosts are exposed sites near the day roost and are often conspicuous because of the presence of large accumulations of guano and discarded insect fragments. In British Columbia, this bat seems to prefer Ponderosa Pines for night roosts; in the United States, it also uses man-made structures, crevices and caves. After feeding, Pallid Bats form clusters at the night roost and if temperatures are cool they become torpid for several hours. Before returning to the day roost they may briefly feed again. Individuals found together at a night roost may occupy separate day roosts.

There are no winter records of Pallid Bats for the province. In the western United States this species is thought to overwinter in the general vicinity of its summer range. Hibernating Pallid Bats have been found in buildings, rock crevices, mine tunnels and caves. Most of these hibernating records are of one or a few individuals—large winter aggregations seem to be rare.

A social bat, this species produces an assortment of vocalizations for communicating in a colony. These calls, most of which are audible to humans, are used in territorial disputes, for directing individuals to a roosting site and in mother-infant communication. Newborns also emit calls that may assist mothers in locating them. Swarming and calling near roosting sites after the bats return from feeding are thought to advertise roost locations to other members of a colony.

The Pallid Bat produces a musky skunk-like odour from glands on the muzzle. There have been no experimental studies to determine the function of this odour—it may be a defensive mechanism for repelling predators.

The only breeding data for the British Columbian population are a nursing female and a male with enlarged testes, both captured on 9 August, which suggests that young are born in July. In the southwestern United States mating takes place from October to December. The gestation period is about nine weeks and young are born in May and June. The Pallid Bat usually bears one or two young, with twins most common; there are a few records of females carrying three or four foetuses. Females are capable of breeding in their first year but yearling females can bear only one young. The age of sexual maturity for males is unknown. Young weigh 3.0 to 3.5 grams at birth; they are undeveloped and their eyes are closed. In four or five weeks they are capable of flight and by eight weeks they attain adult size.

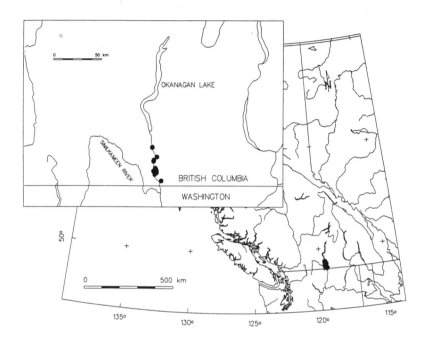

Range

The Pallid Bat ranges from Mexico, throughout the western United States, to Canada where it is confined to the southern Okanagan Valley of British Columbia; there it is found only in a localized area between Okanagan Falls and Osoyoos.

Taxonomy

The British Columbia population belongs to the subspecies *A. p. pallidus*, a variable race that ranges from Mexico to British Columbia.

Remarks

One of the rarest mammals in the province, the Pallid Bat is on the provincial Red List. Only 19 individuals have been captured in British Columbia since its discovery in the province 60 years ago when Ian McTaggart-Cowan and Ken Racey found an individual roosting in a stone pile near Boulder Creek about five kilometres north of Oliver. This was the only evidence for this species in the province until 1974 when a dying Pallid Bat found at Okanagan Falls was submitted to the Federal Agriculture Laboratory to be tested for rabies. In the late 1970s and 1980s, Brock Fenton and Robert Herd captured four Pallid Bats in the vicinity of Vaseux Canyon in the Okanagan. Then, in 1990 and 1991, 12 Pallid Bats were netted in field studies conducted by Mark Brigham and his students. More survey work is needed in the southern dry interior to delimit the range of the Pallid Bat in British Columbia.

Selected References: Bell (1982), Grindal et al. (1991), Hermanson and O'Shea (1983), Orr (1954), Racey (1933).

APPENDIXES

Appendix 1:
Forearm Lengths and Weights (Body Mass) for the 16 Species of Bats in British Columbia

Species	Forearm Length		Weight	
	Mean	Range	Mean	Range
California Myotis (*Myotis californicus*)	32.1	29.4-35.0	4.4	3.3-5.4
Western Small-footed Myotis (*Myotis ciliolabrum*)	31.8	28.8-33.4	4.6	2.8-5.5
Western Long-eared Myotis (*Myotis evotis*)	38.4	36.0-42.0	5.5	4.2-8.6
Keen's Long-eared Myotis (*Myotis keenii*)	36.0	34.2-38.5	5.1	4.0-5.9
Little Brown Myotis (*Myotis lucifugus*)	36.4	33.0-40.3	6.2	6.2-10.4
Northern Long-eared Myotis (*Myotis septentrionalis*)	36.1	34.0-38.0	6.5	5.0-10.0
Fringed Myotis (*Myotis thysanodes*)	42.2	40.0-44.5	7.1	5.4-8.4
Long-legged Myotis (*Myotis volans*)	38.3	34.0-43.0	7.2	5.5-10.0
Yuma Myotis (*Myotis yumanensis*)	34.3	30.0-38.0	6.6	4.0-8.5
Western Red Bat (*Lasiurus blossevillii*)	38.8	34.0-42.0	10.8	7.2-18.5
Hoary Bat (*Lasiurus cinereus*)	54.5	50.3-57.4	31.5	20.1-37.9
Silver-haired Bat (*Lasionycteris noctivagans*)	41.4	39.1-43.9	9.0	5.8-12.4
Big Brown Bat (*Eptesicus fuscus*)	47.5	43.0-52.0	15.2	8.8-21.9
Spotted Bat (*Euderma maculatum*)	50.9	47.9-53.1	17.9	16.2-21.4
Townsend's Big-eared Bat (*Plecotus townsendii*)	42.6	39.9-45.2	8.6	6.0-13.5
Pallid Bat (*Antrozous pallidus*)	52.5	48.0-57.4	17.0	12.0-24.3

Appendix 2:
Dental Formulas for the 16 Species of Bats in British Columbia

Species	Incisors	Canines	Pre-Molars	Molars
California Myotis (*Myotis californicus*)	2/3	1/1	3/3	3/3
Western Small-footed Myotis (*Myotis ciliolabrum*)	2/3	1/1	3/3	3/3
Western Long-eared Myotis (*Myotis evotis*)	2/3	1/1	3/3	3/3
Keen's Long-eared Myotis (*Myotis keenii*)	2/3	1/1	3/3	3/3
Little Brown Myotis (*Myotis lucifugus*)	2/3	1/1	3/3	3/3
Northern Long-eared Myotis (*Myotis septentrionalis*)	2/3	1/1	3/3	3/3
Long-legged Myotis (*Myotis volans*)	2/3	1/1	3/3	3/3
Fringed Myotis (*Myotis thysanodes*)	2/3	1/1	3/3	3/3
Yuma Myotis (*Myotis yumanensis*)	2/3	1/1	3/3	3/3
Western Red Bat (*Lasiurus blossevillii*)	1/3	1/1	2/2	3/3
Hoary Bat (*Lasiurus cinereus*)	1/3	1/1	2/2	3/3
Silver-haired Bat (*Lasionycteris noctivagans*)	2/3	1/1	2/3	3/3
Big Brown Bat (*Eptesicus fuscus*)	2/3	1/1	1/2	3/3
Spotted Bat (*Euderma maculatum*)	2/3	1/1	2/2	3/3
Townsend's Big-eared Bat (*Plecotus townsendii*)	2/3	1/1	2/3	3/3
Pallid Bat (*Antrozous pallidus*)	1/2	1/1	1/2	3/3

Appendix 3:
Scientific Names of Plants and Animals Mentioned in this Book

Plants

Big Sagebrush	*Artemisia tridentata*
Bitter-brush	*Purshia tridentata*
Bunchgrass	*Agropyron spicatum*
Cottonwood	*Populus trichocarpa*
Douglas-fir	*Pseudotsuga menziesii*
Green Ash	*Fraxinus pennsylvanicus*
Ponderosa Pine	*Pinus ponderosa*
Poplar	*Populus tremuloides*
Rabbit Brush	*Chrysothamnus nauseosus*
Salal	*Gaultheria shallon*
Western Hemlock	*Tsuga heterophylla*
Western Red-cedar	*Thuja plicata*

Animals

Marten	*Martes americana*
Pocket Mouse	*Perognathus flavus*
Raccoon	*Procyon lotor*
Red Fox	*Vulpes vulpes*
Spotted Owl	*Strix occidentalis*
Striped Skunk	*Mephitis mephitis*
Western Yellow-bellied Racer	*Coluber mormon*

GLOSSARY

Ambient Temperature: the temperature of the surrounding air.

Biodiversity: the variety of living organisms in an area.

Calcar: a cartilaginous spur that is attached to the heel bone and extends into the tail membrane. This structure is unique to bats.

Echolocation: an orientation system based on generating sounds and listening to their returning echoes to locate obstacles and prey.

Fertilization: the impregnation of the egg by the male sperm cell.

Forage: to hunt for food.

Gestation Period: the length of the pregnancy; the time from fertilization to birth of the foetus.

Gleaner: a bat that can capture prey on the leaves and twigs of vegetation or on the ground.

Hibernaculum: a site where hibernation occurs (pl: hibernacula).

Hibernation: a state of lethargy characterized by a reduction in body temperature and metabolic rate.

Kilohertz: 1000 cycles per second; a unit for measuring the frequency of sound.

Maternity Colony: an aggregation of females and their young.

Metabolic Rate: a measure of the general activity level of an animal; free energy production per unit of body mass.

Nocturnal: active during the night.

Ovulation: maturation and release of the egg before fertilization.

Parturition: birth.

Recurved: curved or bent backwards.

Roost: a daytime retreat or night-time resting place.

Rostrum: the nasal area or the snout of a skull.

Swarming: behaviour associated with nocturnal flights that are made through potential hibernacula by aggregations of bats in late summer or fall.

Thermoregulation: regulation of body temperature by conserving or releasing heat as required.

Torpor: a short-term (daily) state of inactivity achieved by lowering the body temperature and reducing the metabolic rate in order to conserve energy.

Tragus: a thin, cartilaginous structure attached to the base of the ear. This structure is unique to bats.

Transverse Mercator Projection: a map projection based on the principle of a cylinder tangent to the globe along a chosen pair of opposite meridians.

Ultrasonic: sounds above the range of hearing of the human ear (normally above 20 kilohertz).

Universal Transverse Mercator (UTM) co-ordinates: derived from a grid sytem that divides the earth into 60 zones each 6° of longitude. Locations are defined by a zone number and associated east and north geographic co-ordinates.

Vespertilionid Bat: a member of the family Vespertilionidae, a cosmopolitan family of bats with many species inhabiting temperate regions of the Old World and the New World.

Volant: capable of flying.

REFERENCES

General Books on Bats

Barbour, R.W. and W.H. Davis. 1969. *Bats of America*. Lexington: The University Press of Kentucky.

Fenton, M.B. 1983. *Just Bats*. Toronto: University of Toronto Press.

Fenton, M.B. 1992. *Bats*. New York: Facts on File, Inc.

Griffin, D.R. 1958. *Listening in the Dark*. New Haven, Connecticut: Yale University Press.

Hill, J.E. and J.D. Smith. 1984. *Bats: A Natural History*. Austin: University of Texas Press.

Schober, W. 1984. *The Lives of Bats*. New York: Arco Publishing.

Tuttle, M.D. 1988. *America's Neighbourhood Bats*. Austin: University of Texas Press.

van Zyll de Jong, C.G. 1985. *Handbook of Canadian Mammals. 2. Bats*. Ottawa: National Museum of Natural Sciences, National Museums of Canada.

Selected References for General Biology

Barclay, R.M.R., D.W. Thomas and M.B. Fenton. 1980. Comparison of methods used for controlling bats in buildings. *Journal of Wildlife Management* 44:502-506.

Brigham, R.M. and M.B. Fenton. 1987. The effect of roost sealing as a method to control maternity colonies of Big Brown Bats. *Canadian Journal of Public Health* 78:47-50.

Clark, D.R. Jr. 1981. Bats and environmental contaminants: a review. U.S. Department of the Interior, Fish and Wildlife Service *Special Report* 235.

Constantine, D.G. 1988. Health precautions for bat researchers. Pp. 491-528. In *Ecological and Behavioral Methods for the Study of Bats* edited by T.H. Kunz. Washington: Smithsonian Institution Press.

Fenton, M.B. 1982. Echolocation, insect hearing, and the feeding ecology of insectivorous bats. Pp. 261-285. In *Ecology of Bats* edited by T.H. Kunz. New York: Plenum Press.

Fenton, M.B. 1986. Design of bat echolocation calls: implications for foraging ecology and communication. *Mammalia* 50:193-203.

Fullard, J.H., M.B. Fenton, and C.L. Furlonger. 1983. Sensory relationships of moths and bats sampled from two Nearctic sites. *Canadian Journal of Zoology* 61:1752-1757.

Fullard, J.H. 1987. Sensory ecology and neuroethology of moths and bats: interactions in a global perspective. Pp. 244-273. In *Recent Advances in the Study of Bats*, edited by M.B. Fenton, P.A. Racey and J.M.V. Rayner. Cambridge, UK: Cambridge University Press.

Griffin, D.R. 1958. *Listening in the Dark*. New Haven, Connecticut: Yale University Press.

Grindal, S.D., T.S. Collard, R.M. Brigham and R.M.R. Barclay. 1992. The influence of precipitation on reproduction by myotis bats in British Columbia. *American Midland Naturalist* 128: 339-344.

Herd, R.M. and M.B. Fenton. 1983. An electrophoretic, morphological, and ecological investigation of a putative hybrid zone between *Myotis lucifugus* and *Myotis yumanensis* (Chiroptera: Vespertilionidae). *Canadian Journal of Zoology* 61:2029-2050.

Humphrey, S.R. 1975. Nursery roosts and community diversity of Nearctic bats. *Journal of Mammalogy* 56:312-346.

Kunz, T.H. 1982. Roosting ecology. Pp. 1-56. In *Ecology of Bats*, edited by T.H. Kunz. New York: Plenum Press.

Kunz, T.H. 1987. Post-natal growth and energetics of suckling bats. Pp. 395-420. In *Recent Advances in the Study of Bats*, edited by M.B. Fenton, P.A. Racey and J.M.V. Rayner. Cambridge, UK: Cambridge University Press.

Kunz, T.H. (editor). 1988. *Ecological and Behavioral Methods for the Study of Bats*. Washington: Smithsonian Institution Press.

Lyman, C.P. 1970. Thermoregulation and metabolism in bats. Pp. 301-387. In *Biology of Bats*, vol. 1, edited by W.A. Wimsatt. New York: Academic Press.

Meidinger, D. and J. Pojar, editors. 1991. *Ecosystems of British*

Columbia. British Columbia Ministry of Forests, Special
Report 6.

McCracken, G.F. and M.K. Gustin. 1987. Batmom's daily nightmare.
Natural History 96 (10):66-73.

Mitchell-Jones, A.J., A.S. Cooke, I.L. Holroyd and R.E. Stebbings.
1989. Bats and remedial timber treatment chemicals—a review.
Mammalian Review 19:93-110.

Nagorsen, D.W. 1990a. Mammals. Pp. 39-43. In *The Vertebrates of
British Columbia: Scientific and English Names*, edited by R.A.
Cannings and A.P. Harcombe. RBCM Heritage Record 20;
Wildlife Report R24. Victoria: Royal British Columbia Museum
and Wildlife Branch, Ministry of Environment.

Nagorsen, D.W. 1990b. *The Mammals of British Columbia: A
Taxonomic Catalogue*. Memoir 4. Victoria: Royal British
Columbia Museum.

Nagorsen, D.W., A.A. Bryant, D. Kerridge, G. Roberts, A. Roberts
and M.J. Sarell. In press. Winter bat records for British
Columbia. *Northwestern Naturalist*.

Norberg, U.M. 1987. Wing form and flight mode in bats. Pp. 43-56. In
Recent Advances in the Study of Bats, edited by M.B. Fenton, P.A.
Racey and J.M.V. Rayner. Cambridge, UK: Cambridge
University Press.

Norberg, U.M. and J.M.V. Rayner. 1987. Ecological morphology and
flight in bats (Mammalia; Chiroptera): wing adaptations, flight
performance, foraging strategy, and echolocation. *Philosophical
Transactions of the Royal Society of London* B 316:335-427.

Racey, P.A. 1982. Ecology of bat reproduction. Pp. 57-104. In *Ecology
of Bats*, edited by T.H. Kunz. New York: Plenum Press.

Ransome, R. 1990. *The Natural History of Hibernating Bats*. London:
Christopher Helm.

Roeder, K.D. 1967. *Nerve Cells and Insect Behavior*. Cambridge, MA:
Harvard University Press.

Ross, A. 1967. Ecological aspects of the food habits of insectivorous
bats. *Western Foundation of Vertebrate Zoology* 1:205-263.

Thomas, D.W. 1988. The distribution of bats in different ages of
Douglas-fir forests. *Journal of Wildlife Management* 52:619-626.

Thomas, D.W. and S.D. West. 1989. *Sampling Methods for Bats*. U.S.
Department of Agriculture, Forest Service, General Technical
Report PNW-GTR-243.

Thomas, D.W., M.B. Fenton and R.M.R. Barclay. 1979. Social
behavior of the Little Brown Bat, *Myotis lucifugus*. I Mating
behavior. *Behavioral Ecology and Sociobiology* 6; 129-136.

Thomas, D.W., M. Dorais and J.M. Bergeron. 1990. Winter energy
    budgets and cost of arousal for hibernating Little Brown Bats,
    Myotis lucifugus. Journal of Mammalogy 71:475-479.

Tuttle, M.D. 1988. *America's Neighbourhood Bats*. Austin: University
    of Texas Press.

Tuttle, M.D. and D. Stevenson. 1982. Growth and survival of bats.
    Pp. 105-150. In *Ecology of Bats*, edited by T.H. Kunz. New
    York: Plenum Press.

van Zyll de Jong, C.G. 1985. *Handbook of Canadian Mammals. 2. Bats*.
    Ottawa: National Museum of Natural Sciences, National
    Museums of Canada.

Vaughan, T.A. 1970. Adaptations for flight in bats. Pp. 127-143. In
    About Bats, edited by B.H. Slaughter and D.W. Wilson. Dallas:
    Southern Methodist University Press.

Whitaker, J.O. Jr., C. Maser and L.E. Keller. 1977. Food habits of
    bats in western Oregon. *Northwest Science* 51:46-55.

Whitaker, J.O. Jr., C. Maser and S.P. Cross. 1981. Food habits of
    eastern Oregon bats, based on stomach and scat analyses.
    Northwest Science 55:281-292.

Wildlife Branch, Ministry of Environment, Lands and Parks. 1993.
    *Birds, mammals, reptiles and amphibians at risk in British Columbia:
    1993 Red and Blue Lists*. Victoria, B.C.

Selected References for Species Accounts

Aldridge, H. 1986. Manoeuvrability and ecological segregation in the
    Little Brown (*Myotis lucifugus*) and Yuma (*Myotis yumanensis*)
    bats (Chiroptera: Vespertilionidae). *Canadian Journal of Zoology*
    64:1878-1882.

Baker, R.J., J.C. Patton, H.H. Genoways and J.W. Bickham. 1988.
    Genic studies of *Lasiurus* (Chirpotera: Vespertilionidae). The
    Museum, Texas Tech University, *Occasional Papers* 117.

Barclay, R.M.R. 1984. Observations on the migration, ecology, and
    behaviour of bats at Delta Marsh, Manitoba. *Canadian
    Field-Naturalist* 98:331-336.

Barclay, R.M.R. 1985. Long- versus short-range foraging strategies of
    Hoary (*Lasiurus cinereus*) and Silver-haired (*Lasionycteris
    noctivagans*) bats and the consequences for prey selection.
    Canadian Journal of Zoology 63:2507-2515.

Barclay, R.M.R. 1986. The echolocation calls of Hoary (*Lasiurus
    cinereus*) and Silver-haired (*Lasionycteris noctivagans*) bats as

adaptations for long- versus short-range foraging strategies and the consequences for prey selection. *Canadian Journal of Zoology* 64:2700-2705.

Barclay, R.M.R. 1989. The effect of reproductive condition on the foraging behavior of female Hoary Bats, *Lasiurus cinereus*. *Behavioral Ecology and Sociobiology* 24:31-37.

Barclay, R.M.R. 1991. Population structure of temperate zone insectivorous bats in relation to foraging behaviour and energy demand. *Journal of Animal Ecology* 60:165-178.

Barclay, R.M.R., P.A. Faure, and D.R. Farr. 1988. Roosting behavior and roost selection by migrating Silver-haired Bats (*Lasionycteris noctivagans*). *Journal of Mammalogy* 69:821-825.

Bell, G.P. 1982. Behavioral and ecological aspects of gleaning by a desert insectivorous bat, *Antrozous pallidus* (Chiroptera: Vespertilionidae). *Behavioral Ecology and Sociobiology* 10: 217-223.

Brigham, R.M. 1990. Prey selection by Big Brown Bats (*Eptesicus fuscus*) and Common Nighthawks (*Chordeiles minor*). *American Midland Naturalist* 124:73-80.

Brigham, R.M. 1991. Flexibility in foraging and roosting behaviour by the Big Brown Bat (*Eptesicus fuscus*). *Canadian Journal of Zoology* 69:117-121.

Brigham, R.M. and M.B. Fenton. 1986. The influence of roost closure on the roosting and foraging behaviour of *Eptesicus fuscus* (Chiroptera: Vespertilionidae). *Canadian Journal of Zoology* 64:1128-1133.

Brigham, R.M. and M.B. Fenton. 1991. Convergence in foraging strategies by two morphologically and phylogenetically distinct nocturnal and aerial insectivores. *Journal of the Zoological Society of London* 223:475-489.

Brigham, R.M., H.D.J.N. Aldridge and R.L. Mackey. 1992. Variation in habitat use and prey selection by Yuma Bats, *Myotis yumanensis*. *Journal of Mammalogy* 73:640-665.

Caire, W., R.K. Laval, M.L. Laval and R. Clawson. 1979. Notes on the ecology of *Myotis keenii* (Chiroptera: Vespertilionidae) in eastern Missouri. *American Midland Naturalist* 102:404-407.

Constantine, D.G. 1966. Ecological observations on lasiurine bats in Iowa. *Journal of Mammalogy* 47:34-41.

Cowan, I. McT. 1933. Some notes on the hibernation of *Lasionycteris noctivagans*. *Canadian Field-Naturalist* 47:74-75.

Cowan, I. McT. 1942. Notes on the winter occurrence of bats in British Columbia. *Murrelet* 23:61.

Dalquest, W.W. and M.C. Ramage. 1946. Notes on the Long-legged Bat (*Myotis volans*) at Old Fort Tejon and vicinity, California. *Journal of Mammalogy* 27:60-63.

Dalquest, W.W. 1947. Notes on the natural history of the bat, *Myotis yumanensis*, with a description of a new race. *American Midland Naturalist* 38:224-247.

Faure, P.A., J.H. Fullard and R.M.R. Barclay. 1990. The response of tympanate moths to the echolocation calls of a substrate gleaning bat, *Myotis evotis*. *Journal of Comparative Physiology* A 166: 843-849.

Fenton, M.B. 1970. Population studies of *Myotis lucifugus* (Chiroptera: Vespertilionidae) in Ontario. Royal Ontario Museum, *Life Sciences Contribution* 77.

Fenton, M.B., and R.M.R. Barclay. 1980. Myotis lucifugus. *Mammalian Species* 142.

Fenton, M.B., C.G. van Zyll de Jong, G.P. Bell, D.B. Campbell and M. Laplante. 1980. Distribution, parturition dates, and feeding of bats in southcentral British Columbia. *Canadian Field-Naturalist* 94:416-420.

Fenton, M.B., H.G. Merriam and G.L. Holroyd. 1983. Bats of Kootenay, Glacier, and Mount Revelstoke national parks in Canada: identification by echolocation calls, distribution, and biology. *Canadian Journal of Zoology* 61:2503-2508.

Findley, J.S. and C. Jones. 1964. Seasonal distribution of the Hoary Bat. *Journal of Mammalogy* 45:461-470.

Genter, D.L. 1986. Wintering bats of the upper Snake River plain: occurrence in lava tube caves. *Great Basin Naturalist* 46:241-244.

Graham, R.E. 1966. Observations on the roosting habits of the Big-eared Bat, (*Plecotus townsendii*), in California limestone caves. *Cave Notes* 8:17-22.

Grindal, S.D., T.S. Collard and R.M. Brigham. 1991. Evidence for a resident breeding population of Pallid Bats (*Antrozous pallidus*) in British Columbia. Royal British Columbia Museum *Contributions to Natural Science* 14.

Grinnell, H.W. 1918. A synopsis of the bats of California. *University of California Publications in Zoology* 17:223-404.

Herd, R.M. and M.B. Fenton. 1983. A electrophoretic, morphological, and ecological investigation of a putative hybrid zone between *Myotis lucifugus* and *Myotis yumanensis* (Chiroptera: Vespertilionidae). *Canadian Journal of Zoology* 61:2029-2050.

Hermanson, J.W. and T.J. O'Shea. 1983. Antrozous pallidus. *Mammalian Species* 213.

Hickey, M.B.C. and M.B. Fenton. 1990. Foraging by Red Bats (*Lasiurus borealis*): do interspecific chases mean territoriality? *Canadian Journal of Zoology* 68:2477-2482.

Humphrey, S.R. and J.B. Cope. 1976. Population ecology of the Little Brown Bat, *Myotis lucifugus*, in Indiana and north-central Kentucky. American Society of Mammalogists, *Special Publication* 4.

Humphrey, S.R. and T.H. Kunz. 1976. Ecology of a Pleistocene relict, the Western Big-eared Bat (*Plecotus townsendii*), in the southern Great Plains. *Journal of Mammalogy* 57:470-494.

Jobin, L. 1952. New winter records of bats in British Columbia. *Murrelet* 33:42.

Krutzsch, P.H. 1954. Notes on the habits of the bat, *Myotis californicus*. *Journal of Mammalogy* 35:539-545.

Kunz, T.H. 1982. Lasionycteris noctivagans. *Mammalian Species* 172.

Kunz, T.H. and R.A. Martin. 1982. Plecotus townsendii. *Mammalian Species* 175.

Kurta, A. and R.H. Baker. 1990. Eptesicus fuscus. *Mammalian Species* 356.

Leonard, M.L. and M.B. Fenton. 1983. Habitat use by Spotted Bats (*Euderma maculatum*) (Chiroptera: Vespertilionidae): roosting and foraging behaviour. *Canadian Journal of Zoology* 61: 1487-1491.

Manning, R.W., and J.K. Jones Jr. 1989. Myotis evotis. *Mammalian Species* 329.

Martin, R.A. and B.G. Hawkes. 1972. Hibernating bats of the Black Hills of South Dakota. I. Distribution and habitat selection. *Bulletin of the New Jersey Academy of Science* 17:24-30.

Maser, C., B.R. Mate, J.F. Franklin and C.T. Dryness. 1984. *Natural history of Oregon coast mammals*. Special Publication. Museum of Natural History, University of Oregon.

Maslin, T.P. 1938. Fringed-tailed Bat in British Columbia. *Journal of Mammalogy* 19:373.

O'Farrell, M.J. and E.H. Studier. 1973. Reproduction, growth, and development in *Myotis thysanodes* and *M. lucifugus* (Chiroptera: Vespertilionidae). *Ecology* 54:18-30.

O'Farrell, M.J. and E.H. Studier. 1980. Myotis thysanodes. *Mammalian Species* 137.

Orr, R.T. 1954. Natural history of the Pallid Bat, *Antrozous pallidus* (Le Conte). *Proceedings of the California Academy of Sciences* 28(4):165-246.

Parsons, H.J., D.A. Smith and R.F. Whittam. 1986. Maternity colonies of Silver-haired Bats, *Lasionycteris noctivagans*, in Ontario and Saskatchewan. *Journal of Mammalogy* 67:598-600.

Perkins, J.M. and S.P. Cross. 1988. Differential use of some coniferous forest habitats by Hoary and Silver-haired Bats. *Murrelet* 69: 21-24.

Pearson, O.P., M.R. Koford and A.K. Pearson. 1952. Reproduction of the Lump-nosed Bat (*Corynorhinus rafinesquei*) in California. *Journal of Mammalogy* 33:273-320.

Poché, R.M. 1981. *Ecology of the Spotted Bat* (Euderma maculatum) *in southwest Utah*. Utah Division of Wildlife Resources, Publication 81-1.

Racey, K. 1933. Pacific Pallid Bat in British Columbia. *Murrelet* 14:18.

Ross, A. 1967. Ecological aspects of the food habits of insectivorous bats. *Western Foundation of Vertebrate Zoology* 1:205-263.

Saunders, M.B. 1989. Resource partitioning between Little Brown Bats (*Myotis lucifugus*) and Long-legged Bats (*Myotis volans*) in southern Alberta. M.Sc. thesis, University of Calgary.

Schowalter, D.B. 1980. Swarming, reproduction, and early hibernation of *Myotis lucifugus* and *M. volans* in Alberta, Canada. *Journal of Mammalogy* 61:350-354.

Schowalter, D.B., W.J. Dorward and J.R. Gunson. 1978. Seasonal occurrence of Silver-haired Bats (*Lasionycteris noctivagans*) in Alberta and British Columbia. *Canadian Field-Naturalist* 92: 288-291.

Schowalter, D.B. and J.R. Gunson. 1979. Reproductive biology of the Big Brown Bat (*Eptesicus fuscus*) in Alberta. *Canadian Field-Naturalist* 93:48-54.

Schowalter, D.B., J.R. Gunson and L.D. Harder. 1979. Life history characteristics of Little Brown Bats (*Myotis lucifugus*) in Alberta. *Canadian Field-Naturalist* 93:243-251.

Shump, K.A. Jr. and A.U. Shump. 1982a. Lasiurus borealis. *Mammalian Species* 183.

Shump, K.A. Jr. and A.U. Shump. 1982b. Lasiurus cinereus. *Mammalian Species* 185.

Tuttle, M.D. and L.R. Heaney. 1974. Maternity habits of *Myotis leibii* in South Dakota. *Bulletin of Southern California Academy of Sciences* 73 (2):80-83.

van Zyll de Jong, C.G. 1979. Distribution and sytematic relationships of long-eared *Myotis* in western Canada. *Canadian Journal of Zoology* 57:987-994.

van Zyll de Jong, C.G., M.B. Fenton and J.G. Woods. 1980. Occurrences of *Myotis californicus* at Revelstoke and a second record of *Myotis septentrionalis* from British Columbia. *Canadian Field-Naturalist* 94:455-456.

Wai-Ping, V. and M.B. Fenton. 1989. Ecology of the Spotted Bat (*Euderma maculatum*) roosting and foraging behavior. *Journal of Mammalogy* 70:617-622.

Warner, R.M. and N.J. Czaplewski. 1984. Myotis volans. *Mammalian Species* 224.

Watkins, L.C. 1977. Euderma maculatum. *Mammalian Species* 77.

Woodsworth, G.C. 1981. Spatial partitioning by two species of sympatric bats, *Myotis californicus* and *Myotis leibii*. M.Sc. thesis, Carleton University, Ottawa.

Woodsworth, G.C., G.P. Bell and M.B. Fenton. 1981. Observations of the echolocation, feeding, behaviour, and habitat use of *Euderma maculatum* (Chirpotera: Vespertilionidae) in southcentral British Columbia. *Canadian Journal of Zoology* 59:1099-1102.

ACKNOWLEDGMENTS

We are indebted to the many individuals who made this book possible. Robert Barclay, Cathy Koehler, Joseph Cebek, Tod Collard, Paul Faure, Scott Grindal, Brian Hickey, Dorothea Krull, Susan Lewis, Juan Carlos Morales, Martin Obrist and Matthew Saunders generously provided various information, some unpublished, from their research on bats. We thank Anna Roberts and Gina Roberts for their bat observations from the Williams Lake region, Chris Harris and Barry Milligan for unpublished data on the Yuma Myotis colony at Squilax, Mitchell Firman and Mike Getty for unpublished data from the Keen's Long-eared Myotis survey in 1991, David Kerridge for unpublished data on Vancouver Island bats, Mike Sarell for his various unpublished bat observations from the province, Susan Holroyd and Lisa Merk for unpublished data from their 1992 surveys, and Kelly Chapman and Kathleen McGuinness for unpublished data on the Pallid Bat study in the Okanagan. Anne Brigham, Andrew Bryant, Wayne Campbell, Alan Dundas, Chris Dodd, Rick Howie, Kelley Kissner, Dave Low, Herb Matthews, Bryan Webster, Olivia Whitwell and Kevin Zurowski provided various information on bats in British Columbia and/or assisted in the field. Brian Lawhead kindly provided his unpublished manuscript on bat distributions in Alaska. C. G. van Zyll de Jong verified the identifications of problematic museum specimens of the long-eared *Myotis* group.

Elizabeth Taylor extracted all Royal British Columbia Museum specimen records stored in CHIN (Canadian Heritage Inventory Network) and assisted with setting up a microcomputer data base for all British Columbia bat records. R. Yorke Edwards calculated UTM co-ordinates for the many locality records and diligently tracked down many obscure historical museum records; he also commented on a preliminary draft of the book. The Environmental Youth Corps, under direction of the Wildlife Branch of the Ministry of Environment, entered museum records from the National Museum of Canada into our data base.

We thank the following institutions for data from British Columbian bat specimens housed in their collections: American Museum of Natural History, New York; California Academy of Sciences, San Francisco; Carnegie Museum of Natural History, Pittsburgh; Cowan Vertebrate Museum, University of British Columbia, Vancouver; Field Museum of Natural History, Chicago; Museum of Comparative Zoology, Harvard University, Cambridge; Museum of Vertebrate Zoology, University of California, Berkeley; Museum of Zoology, University of Michigan, Ann Arbor; National Museum of Canada, Ottawa; Philadelphia Academy of Sciences, Philadelphia; Royal Ontario Museum, Toronto; United States National Museum, Washington, D.C.; University of Kansas, Lawrence. The California Academy of Sciences and the National Museum of Canada also loaned specimens for artistic reference material. James Patton (University of California) and Susan Woodward (Royal Ontario Museum) provided additional measurements from material in their collections.

Research for this book was supported by the Royal British Columbia Museum, the Wildlife Branch of the Ministry of Environment, and an operating grant from the Natural Sciences and Engineering Research Council (NSERC) to RMB.

We are indebted to Brock Fenton and J. Knox Jones Jr. for their thorough and constructive reviews.

INDEX

The Mammals of British Columbia

Six volumes:

Edited and designed by Gerry Truscott, RBCM
Cover design by Gerald Luxton, RBCM
Illustrations by Michael Hames, with enhancements by Gerald Luxton
Typeset in Plantin 10/13 by The Typeworks, Vancouver, B.C.
Printed and bound in Canada by D. W. Friesen & Sons Ltd.